SPACIOUS SKIES

SPACIOUS
SKIES

Richard Scorer
& Arjen Verkaik

DAVID & CHARLES
Newton Abbot   London

This book is dedicated to all those who love and respect the Earth in humility. The sky is its life-blood and the clouds its pulsating heart, the wind its spirit. With all the other species we share the opportunities for living in the ocean and atmosphere which nurture us all, under the sun; draw out new moods; cleanse and freshen our surroundings: our collective home!

Page 2 (24/8/87, 09:10 AM, 2)
Cyclones moving up the 20°W meridian

Pages 8/9 The sky encloses all forms of life, but changes more rapidly than any . . .

AUTHORS' NOTE
The picture captions contain the following information where available and appropriate:
    Satellite pictures (Chapters 2, 3, 7–12): day/month/year, time (GMT), satellite channel.
    Others (Chapters 4–6 and 13): day/month/year, local time, direction of view, angle of lens.

British Library Cataloguing in Publication Data

Scorer, R. S. (Richard Segar) *1919–*
    Spacious skies
    1. Clouds. Dynamics
    I. Title   II. Verkaik, Arjen
    551.57'6

    ISBN 0-7153-9139-9

Printed in Portugal by Resopal
for David & Charles Publishers plc
Brunel House   Newton Abbot   Devon

CONTENTS

PREFACE

'The most beautiful experience we can have is the mysterious. It is the fundamental emotion that stands at the cradle of true art and true science. Whoever does not know it and can no longer wonder, no longer marvel, is as good as dead, and his eyes are dimmed.'

Albert Einstein, *The World As I See It*, in Living Philosophies, 1931

This book is the work of two enthusiasts whose passion for the sky infuses their observations and explorations with the lifeblood of science – a sense of wonder. Although Richard S. Scorer and Arjen Verkaik have ventured into meteorology from very different directions, their unflagging enthusiasm for discovering the beauties and mysteries the weather presents for us has prepared the ground for a very fruitful collaboration.

For Professor Scorer the study of sky has been an obsessive pursuit since the early 1950s, when he and Frank Ludlam did forecasting for gliding competitions and studied the motions of air using gliders. He extended his fieldwork into hiking expeditions to watch the rapid cloud changes at sunrise and sunset, and soon found himself carrying his camera wherever he went, by air, by sea, or on foot.

Although the face of professional meteorology was becoming increasingly dominated by giant computers and expensive apparatus he was fortunate enough to find himself among colleagues at the Imperial College who appreciated the value of keeping their eyes on the real thing – the great outdoors. And he found that, although important scientific discoveries were being made using elaborate equipment, discoveries of equal importance are often made simply by watching – looking for those puzzles and paradoxes which the sky presents to challenge the limits of our understanding.

Professor Scorer's investigations have carried him through studies of waves, cumulus, air pollution, turbulence, and now into satellite photographs, his curiosity and expertise increasing with each foray into the wonders of our atmosphere. His studies have always been occasions of discovery. He has looked, not for confirmation of what he thought he already knew, but for fresh discoveries, things which hadn't previously been appreciated or recognised.

While Richard Scorer was adventuring into the delights of the atmosphere through glider studies, Arjen Verkaik, a young boy in Canada, was developing a consuming passion for the sky. He kept a detailed weather log, spending many blissful hours studying the ever-changing skies of northern Ontario. He was fascinated by every detail, from the behaviour of Sudbury's giant smoke plumes to the expansive cirrus feathers that heralded approaching weather systems. He became well known to the weather service later in Toronto, peppering them with questions to feed his insatiable curiosity about what he experienced.

As he grew older his passion for the sky grew with him, first leading him to take up a camera to record the wonders of weather, and then leading him to study physics with an eye to entering meteorology. But the formal pursuit of a degree in meteorology lost out to the growing recognition he was receiving for his photographic documentation of weather. Building on the observational skills he had developed in skywatching, he fast became known for both the beauty and the scientific value he offered in his photography. As his work developed he found that his photographs raised questions for which there were often no satisfactory answers. What had begun as photographic documentation of the sky matured into an excercise in discovery. Working as a team with his wife, he built his passion into an enormous collection of sky photographs and time-lapse film. They have travelled throughout North America (and anywhere else chance took them), chasing the sky to record its wonders, both for scientific scrutiny and aesthetic appreciation.

When Arjen Verkaik went to Britain for a conference in 1984, people repeatedly suggested that he and Professor Scorer should get together. It was a brief meeting, but full of excitement as both men discovered their common passion for exploring the science of the atmosphere. Through many letters over the next years, and with the encouragement of colleagues in the meteorological community, this book was born.

It is never a simple matter to collaborate on a book, and doing so from opposite sides of the Atlantic Ocean has its own special difficulties. But it has been an exciting and challenging experience. As different as Arjen Verkaik and Richard Scorer are in their styles and backgrounds, their shared enthusiasm for exploring and understanding the mysteries behind the sky's artistry has created a bond which speaks out from the pages of this volume. It has been very enjoyable watching their kinship grow with this book. As Francis Bacon wrote:

'. . . all knowledge and wonder (which is the seed of knowledge) is an impression of pleasure in itself . . .'

It is indeed a pleasure to explore the sky through these photographs and satellite pictures, discovering the swirling beauty and harmony of the atmosphere – the bloodstream of our blue planet. Let us see, then, what questions the sky poses for us, from our vantage point on Earth, and from our eyes in space.

Jerrine Craig Verkaik

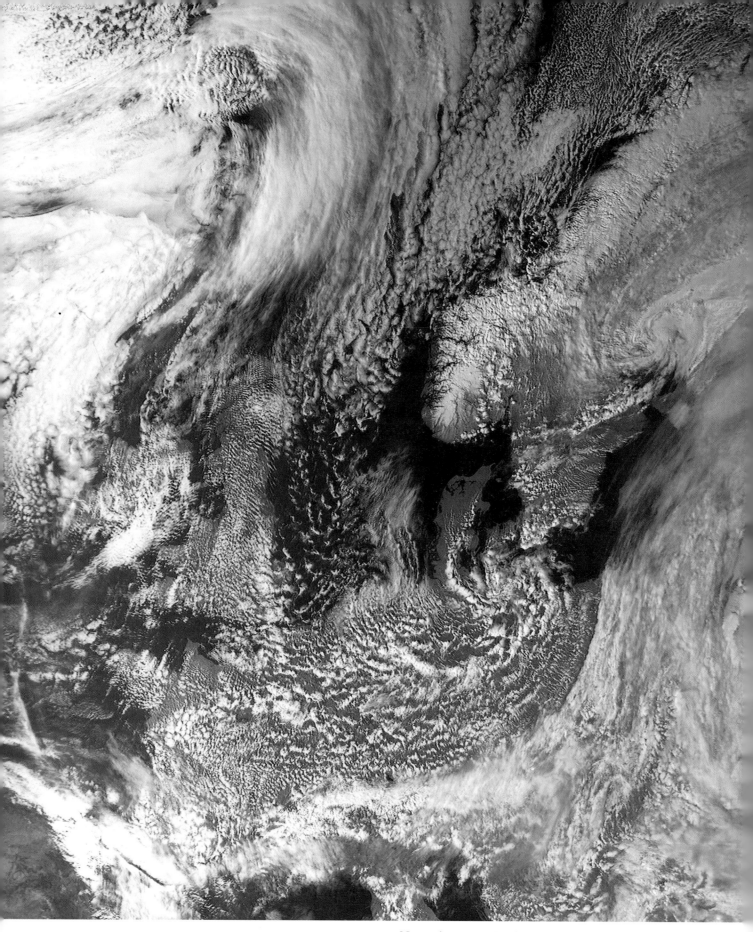

▲ 1.1 (8/4/80, 14:11 PM, 2)

1.1 A cold snap in spring. The cold front behind a cyclone which crossed Scandinavia into the Baltic Sea has reached the Alps and the Pyrenees. Denmark is protected by the Norwegian mountains from the full blast of air from the Greenland Sea, but over Britain the air is more stable as a warm front approaches from the west, and so the country is covered with lee waves.

1
SKYWATCHING

'The sky is the daily bread of the eyes.' (*Emerson*)

WISDOM IN THE WIND

From the lifelike forms of cumulus lumps to the delicate patchwork of an altocumulus sheet to the expansive purity of a cirrus veil, clouds are our constant companions. The sky is a place of beauty and comfort. Within its familiar embrace we share a common experience with other peoples, times and places. Sky is home to our strongest emotions, deepest thoughts, to yesterday's warm memories and tomorrow's wildest dreams.

Long before my teen years, I had already become a skywatcher. The fifteen-minute walk from home to school over a rather high hill provided me with many opportunities to watch the clouds over the region. On some lunch hours, I would become so distracted by the shadows of lazy cumulus that I forgot the time and had to race home for a five-minute sandwich. I remember the giant smelter stack to the west, belching huge gobs of brown, sulphurous smoke over the town. On windy days they curled down and dragged along the ground, but on calm days the plume went straight up and was crowned with a brilliant cumulus cloud that changed constantly. I used to lie in bed on those cold, windy winter nights listening to the hydro lines hum and watching their shadows swing to and fro on the ceiling. I would judge the wind's strength by the extent of swing and usually fell asleep waiting for that next, bigger gust. And the next morning, soft snows had been moulded into shapely drifts with delicate details and a thick crust you could walk on. I remember the surprise of sudden squalls, strange cloud forms, sparkling trees after an overnight ice storm. I would watch all day for the first snowflakes to fall from a stratocumulus sheet on that first real cold wave, or sit on a hill staring into overcast skies for that first distant flicker of lightning the forecast had promised me. I soon became a weather-nut with all the peculiar quirks one might associate with such an eccentric habit. While others listened to popular music on the radio, I would tune in to static to follow the discharges of faraway storms. I kept a log for years using simple picture symbols for every aspect of the weather, and when doodling, drew simulated weather maps for imaginary stations in a fantasy climate.

The stories are as endless as life itself. I'm sure my respect for nature helped shape my early attitudes but all children are curious about the world around them. And after all, isn't the sky a natural place to look? Weather shaped my personality as well. In its endless variety and complexity, it allowed me to express myself more freely and fully than sports or school or even friends could. In the sky, there was always something new to see, to ponder, to discover, and it wasn't very long before I had weather flowing through my veins.

Every region has special skies that repeat there only. They may be rare generally but be taken totally for granted in that local area. When we think we have mastered a full understanding of weather events it is time to consider the area on the far side of the hill, across the lake, other latitudes, other countries, because those places have often very different skies that they, too, thought were typical and average. The complexity never fails to amaze me. The similarities bring us together under one roof while the differences teach us to observe and appreciate the smaller subtleties that bring richness and flavour to our lives. Sky does teach us about itself, but even more importantly, it brings forth and magnifies many of the mysteries in ourselves.

SKY-SMARTS

The sky is the stage on which a perpetual weather drama is played out. To understand the play we must become familiar with the characters and with the context in which they perform. The characters are clouds, the visible tracers of motion and change in the atmosphere. They behave flawlessly in a vast three-dimensional fluid world that is full of variability, constantly subjected to external influences, but always in a context of perfect harmony in response to the laws of physics.

You don't need to read a veterinary text to figure out how your pet behaves. With a little understanding and a lot of patient watching, all the relationships and reasons become apparent on their own. It's the same with clouds. Those reasons and relationships are always present, always directly linking that which can be seen with a set of discrete conditions and mechanisms. Much of the complexity results from the number of variables present and the way in which their relative contributions can vary over time. But despite this, the dedicated skywatcher has the ability to develop an intuitive understanding, a sort of 'sky-smarts' which provides many pleasures and benefits and lays a receptive foundation for a closer look at meteorology.

Many people believe that clouds merely 'move in' from somewhere, stay around for a while, then 'move out' again. This is not surprising, since with an occasional glance out a window or while briefly outside one sees a sky changing in large, obvious steps. This notion is probably reinforced by the evening weather forecast which focuses on large-scale cloud systems that appear to move en masse across the region. Although the finer points of change are lost to a busy lifestyle, it is these very changes – gradual, subtle, continuous – that influence a cloud's appearance the most and offer us the best clues to the atmosphere's characteristics and behaviour. Forward motion can tell us about winds at various altitudes but formation, evaporation, element grouping, etc, bring many other factors to our attention and provide the blueprint for a complete picture of atmospheric processes.

What Clouds Can Tell Us

Clouds are useful as both short and longer term weather predictors. On the longer term (hours to days) they form the components of synoptic-scale systems and their behaviour can be used to interpret or refine the more general daily weather forecasts. Accelerated cloud development, especially in the mid-level, may indicate a more rapid approach of a front or trough. By comparison, persistent thinning of mid and high cloud layers despite thicker cloud upwind and an overcast forecast suggest a 'resistance' to the forward progression of a system due either to slowing or to erosion under reinforced subsidence. And these changes can become even more meaningful when we become familiar with current and predicted winds at various heights, the track the front or pressure centre will take, the modifying effects of local or topographical features, and other bits of weather information. The framework of knowledge we accumulate is then stimulated by each new cloud circumstance to offer up a reliable indicator of future sky events.

The most obvious thing clouds portray is air motion, which can vary widely in speed and direction with height. Higher cloud layers will move with the prevailing winds aloft that carry weather systems, but lower clouds reflect surface pressure patterns and local influences. Distinguishing these two often independent contributions to the day's weather helps to give meaning to the changes that occur. For example, a thickening, low stratocumulus deck might result when the surface wind pattern changes or surface air becomes moister but has little to do with conditions aloft which might remain fair and dry. In such cases, a little extra knowledge of the hows and whys of atmospheric motion helps to qualify our interpretations of sky changes.

Interpreting cloud motion is complicated by the cloud's height. High cirrus can be seen at the horizon with good visibility and may be 100–200km (62–124 miles) away. They may approach in an hour on a strong upper jet or may take all day in light winds aloft. The same relatively light winds can make very low stratus or stratocumulus look like it is speeding by overhead. To know exactly what it is that we are seeing becomes a matter of interpretation, first of cloud types and their reasons, then of their heights and relationships to air motion. From a high place you can watch cloud shadows move across the land, to help you match features with facts. Clouds seen from the sun's point of view are all practically on the Earth's surface so shadows will portray a cloud's size and motion almost exactly. But then, there are other factors which can influence our perception of things too. The sun's

position in the sky will affect the angle of illumination and thus the extent and position of shading, while the presence of multiple layers can fool even the sharpest observer. The ultimate test of our skills comes when we are able to distinguish straightforward cloud movement from the actual changes in shape, size and structure that are always under way as well.

A Closer Look at Clouds Changing

Distinguishing cloud development and change from cloud movement is the first great challenge which enables us to perceive sky events in a new light (the second being a detailed appreciation of how and why clouds change). Whole sheets of cloud can form or dissolve to give the illusion of movement. But rather than seeing air motion we are witnessing the shifting of one or more factors contributing to the cloud's presence. An interesting aspect of weather is that many of its components are not visible themselves, but are observable by inference. An extreme example might be a standing wave to the lee of a mountain in clear sky, made visible later when increasing moisture permits a wave cloud to form in the already present crest. A common case is a trough rolling forward at a speed different from air motion, thus causing clouds to form or thicken as its effect on the atmosphere approaches. In this way, the system propagates forward, constantly creating new clouds at the leading edge while leaving other clouds behind to evaporate in its wake. Similarly, an upper ridge can overtake existing clouds and dissolve them or inhibit further growth as subsidence and warming aloft dry and stabilise the air in its path. The dedicated skywatcher acquires a 'feel' for these things and soon sees the atmosphere as a single, evolving entity whose individual manifestations are all meaningfully connected to each other.

The smaller scale of single cloud elements, patches, or parts thereof, is a world within a world. Here, changes are often slow or subtle and demand both our patience and attention to detail. Elements are forming, evaporating, or shearing in various ways. There may be particle fallout, transformation of water to ice, and a variety of patterns that emerge then fade again. Keeping track of the three-dimensional air motions by applying a mental outline of imaginary flow lines to the scene will help to structure it so that the changing features are less confusing.

Once features are recognised, it is time to sense them in the wider context of their evolution. We can think of this as a combination of various effects all evolving within an ordered process. If we know or discover the process, we can infer the cloud's behaviour and see where its present state fits on the continuum. To achieve this often requires an extra commitment to watch every changing detail from beginning to end. Even when no obvious effects are visible, things are happening up there because all clouds undergo some degree of transformation. An old cumulus, for instance, may look the same for a while even though internally its droplet size distribution is shifting dramatically. Watching the full history of an event unfold in the sky brings clouds to life and enables us to catch those brief, rare, transitional moments in the view.

To enjoy clouds you need to be a part of their world,

flow with them, feel their history unfold. You then begin to see things that were hidden by the unfamiliar ways of an all-too-familiar sight. You begin to read the lines between the lines and find a message clearer than fact in thoughts and emotions echoing Nature's handiwork.

It's always fun to watch clouds for the exceptional scenes, which might be extreme events, rare occurrences, or very brief transitional moments. Rare events such as tornadoes or sundogs simply don't happen very often. Extreme events (those that depict ordinary things happening in an extreme way or situation) such as very high-based cumulus, low cirrus, or large expanses of similar elements, offer unusual views of familiar cloud types and are also relatively rare. The transitional effects occur regularly but are seldom seen by most people because they last only minutes and occur between more common, less inspiring cloud forms. However, they yield some of the most beautiful, most intriguing views possible in the sky and are well worth seeking out and waiting for.

There are so many good examples of transitional moments, those visually pleasing peaks on the continuum of change, that only a few can be noted here. Most will be elaborated on in other parts of this book. Many occur around convective systems due to brief changes in air motion; these include scud, roll clouds and mamma. Iridescence is evident only in the brief water stage of uniform cirrocumulus or high altocumulus patches. Wave clouds and pileus achieve brief moments of perfect streamlining or symmetry while billows and regular element arrangements pass through a phase of optimal clarity and sharp detail. Altocumulus floccus litters the sky with tiny puffs only briefly when the original sheet breaks up and evaporates. Even the familiar towering cumulus cloud has a moment when its boiling top is crisp and sharp, before it softens again. The ultimate one, of course, is that moment of absolute harmony when we experience the magic of sky in our souls.

A Different Way of Seeing

There is another way to look at skywatching, a way that depends not on the subject itself but rather on our methods and approaches to it. We could call this 'skywatching with discerning eyes and tinted vision' because the results depend on careful attention to detail guided by a predominant viewpoint. The actual method involves scanning the sky for details that seem different; certain shapes, shading and colour, patterns, etc that stand out or catch our curiosity. The 'tinting' is a basic state of mind governing how we interpret what we see, and can take three forms. In the active state we employ an analysis of fine details, features, and element associations to determine what's going on. A second state, mostly passive, relies on the matching of memory and experience to a given scene for its interpretation. This approach, although quite dependable, is under-used in our society. The last state is the single most overlooked means of acquiring and assessing knowledge, the intuitive one. This approach is entirely passive and uses highly sophisticated pattern recognition founded on previous exposures to sky situations and accompanying associations in memory. It is hard to analyse or prove but

yields remarkable results when exposed to a complex, evolving process such as skywatching. It enables a person to become 'tuned-in' to weather and clouds in a special, personal way.

A commitment of time and devotion to sky opens up the subject of meteorology in a unique way to make sense of the apparent chaos and mystery of clouds and weather. As our knowledge expands, the horizons of our perception of sky also widen, until the whole atmosphere comes into focus. The benefits go beyond information, beyond amateur forecasting, beyond the obvious pleasures of a beautiful scene, into another dimension of satisfaction.

RECLAIMING OUR HERITAGE

Sky and weather are not so abstract that we can't develop a relationship to them. In the obvious sense this means a sharing of information wherein sky passes on clues about weather wisdom and we assess them in the light of scientific fact and personal experience. An ensuing familiarity with sky then gives us many other, less obvious, practical benefits like a keener sense of awareness of nature and a growing recognition of how inseparable we really are from our environment.

Watching the sky raises our appreciation for science in general by bringing it directly into our everyday experiences. As a result, we begin to appreciate clouds for their own sake. Those foolish notions that 'good weather' means clear days and blue skies are soon dispelled and replaced with a more realistic response to weather. Can you imagine all those people who step out of bed and permit their thoughts and moods to slump just because the day is overcast! Even the seemingly dull overcast sky has its special offering of windswept fragments, scented moistness, soothing greys and quiet light to share with us. Discovering our environment and getting to know it better makes us want to see the same familiar skies over and over, but always in search of new experiences and the surprises they harbour.

Because of the complexity of sky, it is impossible to apply many rules to being a critical sky observer. Instead, it is better to rely on our other skills to provide a gradual, thorough understanding of this subject. Despite rigorous methods and a technological garb, science has no monopoly on truth or fact. We ought to shrug off the intimidation of expertise and the convenience of ready-made answers and become the authors of more personal and relevant associations with the world around us. This by no means implies the abandonment of present scientific methods that test and prove our notions, but rather their supplementation with an alternative way of assessing information. Through the application of pattern recognition, relational thinking, and an interactive awareness as they pertain to sky, we develop our intuitive abilities. These abilities are better able to cope with the meaning of things in the larger context of life, and make each one of us participants in the advancements of scientific knowledge.

Our relationship to sky is one-to-one, and from it we learn to see things differently. You can't be a good observer of sky, or anything else for that matter, unless you give yourself to the subject, open up to it, let it flow through you. This is much like spiritually centered living in that you allow sky to live within and express itself through you. To really see things in their fullness we need to become a part of what we're looking at.

Sky is a convenient, readily available way to enjoy ourselves. And in the process, we become less detached from our surroundings. We begin to participate and appreciate the things we are exposed to rather than passively enduring their presence. In doing so, we take another step towards the reclamation of a nearly lost heritage in which personal responsibilities have been supplanted by collective indifference.

METAMORPHOSIS

Like the elusive images of dreams, clouds display their twinkling beauty in passing instants, flashes of vapoury charm. Molecules rise, collect, reflect, giving impressions of substance and shape. We identify, respond, slip into an ocean of thoughts, feast on the richness of sky.

The sky is a workspace for the imagination. When we were young, we spent hours gazing at clouds in search of familiar sights and shapes. Those activities were more than idle daydreams, more than the frivolities of young and foolish minds. They represented our search for meaning, for belonging, a reaching out to strange and wonderful sights, a step on the road to discovery. Now, more than ever, we need the opportunity to let thoughts roam free and allow that vast reservoir of self-expression to attain its full potential. Experiencing sky releases us from the constraints of purely linear thinking, enabling the rich undercurrent of the mind to surface and express itself through the imagination.

Sky functions as a catalyst for the expression of the spirit and serves as a metaphor for the transformation of self. The continuous metamorphosis of sky encourages us to press on, reach out, open up, inspiring our personal inclinations toward transformation. It is a process of renewal, and when we become a sky reader, we let the words of its infinite volumes open new dimensions in our appreciation of life and ourselves.

STANDING AT THE DAWN

When it comes to knowing the sky fully, despite all the physics, reading and examples, experience is still the best teacher. It enables us to feel the changes and let our thoughts and emotions go with them. The ultimate experience is the 'chase' in which we no longer let sky pass us by but travel with it in search of rarer views and deeper thrills. Clouds are the signatures of the physics of the atmosphere. They deserve to be understood but also to be viewed in their larger context, both in the physical sense of the entire atmosphere and in the philosophical sense of man's relationship to nature. Sadly, much of science today compartmentalises information and sees the subject as separate from the world and not an intrinsic part of it. Specialisation fragments it further and interconnections become lost or unclear. Meteorology is no exception. In some ways it has a greater responsibility to change this trend, because weather is the common ground for everyone's activities, the perfect communicator of shared experiences, a resource lying dormant in the fields of progress. Today, little credence is given to simply watching, which is considered 'too subjective' to be of much use as a basis for understanding. However, the results speak for themselves. The benefits are countless, but even if they only lead to personal satisfaction and pleasure, they're worth it all.

The universality and timelessness of weather gives us a link to history and the passing of time. Napier Shaw called weather: 'This pageant which has been transforming yesterday into tomorrow ever since the world began.' Our experiences with sky serve as a backdrop and reference point for our other daily activities. The stresses imposed by extreme weather circumstances bring into conscious realisation our interconnectedness to all, forcing us to pay more attention to the things which affect us. We remember special weather events and weave them in among our other memories. This linking is not a deliberate process but, rather, a subconscious one, not unlike the way in which a scent can trigger a specific recollection. If we generalise this idea, we could imagine the daily weather events as the 'string' that joins the beads of day-to-day memories and experiences, a common thread of life. We usually remember the 'beads' but don't necessarily recognise the 'string'. Yet it is this very linking which roots us securely in the progression of time and our relationship to all aspects of life.

The atmosphere is an amazing part of our world. Its largeness, grandeur and majesty parade perpetually before our eyes. Its delicate beauty and infallible ways represent a mastery of perfect harmony that accommodates and forgives our thoughtless impositions. It is the skin of the Earth, a mere membrane of air that is, surely, 'the finest collaboration in all of nature' (Lewis Thomas).

2
EARTHWATCHING

EYES IN THE SKY

The bloodstream of the biosphere is exquisitely evident when viewed from space. In our generation it has become possible to see the Earth from outside. We can watch our local bit of sky every day, but our imagination yearns to go beyond. Art and literature show that wishful thoughts of riding upon the clouds have been common throughout the ages. The sky is where heaven used to be, but now we can reach even higher; transport aircraft take us into the stratosphere, and with satellites we have eyes in orbit.

In Chapter 7 we give a brief description of how we obtain satellite pictures, and what they mean. Here, a few scenes demonstrate the widened view to which we now have access. We are not interested merely in illustrating things we knew about before satellites broadened our perspective; we are looking for new patterns which had not been imagined before the advent of satellite photography.

All life is in the air, the soil, or the ocean; ours is in the air. We now look at the atmosphere, which is one of the three realms of the biosphere and was moulded by life from the lifeless geology of the young Earth into the semi-transparent mixture of pure nitrogen, pure oxygen, pure water vapour, and pure carbon dioxide which now makes our planet the jewel of the solar system.

▲ 2.2 (4/2/80, 14:17 PM, 2)

A REMOTE ISLAND

The southern oceans are large and often stormy. A remote island protrudes here and there from the sea into the gales, and stirs clouds into shapes like a ship-wake, or the wave pattern made by a duck on a pond.

These islands are the home of the great wandering albatross, whose territory surrounds Antarctica. Each year they circle that continent several times, gliding almost without effort, continuously for days on end, to return home now and then to meet their mates. This feat of navigation is helped by their ability to spot their islands by the shapes they impose on the clouds.

Even in the vast expanse of the southern oceans each locality has its own sky, different from the others like the 'pays' of France which a geographer sees as the unique product of soil and topography. This picture is of the sky of Crozet Island (46 deg S, 51 deg E), whose wave pattern in the clouds reaches 200km (125 miles) to the south and 500km (311 miles) to the east, callings its wandering residents home from the very open sea.

2.1 (14/4/86, 11:09 AM, CZ5) ▶

A ROBE OF WAVES

It is not very difficult to explain most of the details seen in wave patterns behind mountains. This piece of dress designing for the Mediterranean island of Corsica not only has a twist in the main train, but has a second system at right angles to it. Not content with that, a set of tiny billows is woven across the foot of the robe. The island of Sicily, to the south, seems to be emerging from a region of more nebulous cloud, while mainland Italy hides under a less ambitiously decorated toga.

ARCTIC WASTES

The far north is a region of great importance to our weather. In contrast to the Antarctic continent it is a frozen ocean surrounded by land and every winter makes it inhospitable to all except animals adapted to life there.

Birds escape the winter darkness by migration, but we can glimpse the icebound desert from aircraft travelling between the rich industrial centres of Europe and eastern Asia or western America. Or we may take the satellite and inspect the great northern island of Greenland. It is a snow plateau, steep walled, sloping from about 3,000m (9,843ft) in the corner nearest to Europe down to about 2,000m (6,562ft) in the west, where it looks across the Davis Strait and Baffin Bay to the Canadian Islands.

The plateau is impassable to the lower level air, which lies beneath an inversion (the name used for a stable layer which keeps the cold surface air beneath warmer air above). Consequently the plateau is very sunny in summer and never sees shower clouds or cold fronts, and the higher clouds of warm fronts which do invade are cool shadows of their name.

◀ 2.3 (1/7/77, 12:17 PM, VIS)

▼ 2.4 (14/4/83, 16:38 PM, 2)

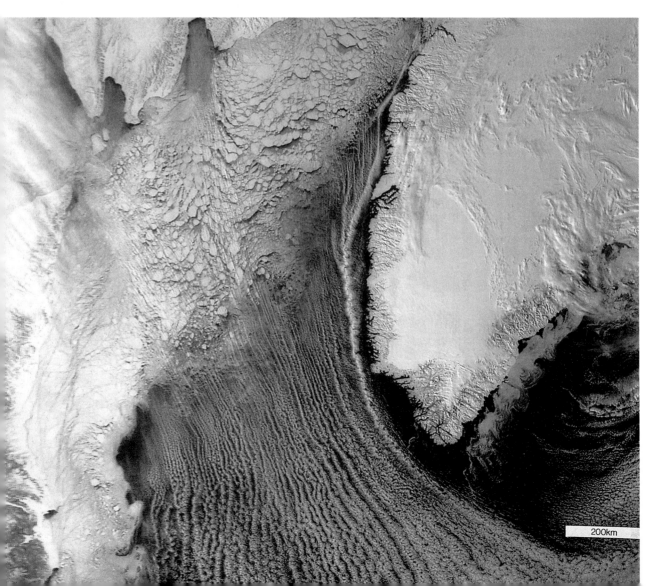

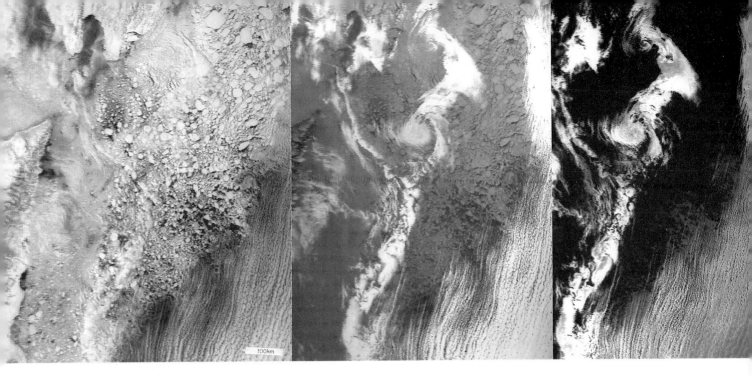

The white snow reflects most of the sunshine and its own radiation into space makes it a very cold surface. The air close to the surface is thereby cooled, and it runs off the plateau like rain off a roof. The continual flux of this cold air across the west coast holds off the westerly winds, which are diverted southwards. With our eyes in space the power of this phenomenon is clearly visible, and the repeated occurence of scenes like this one demonstrates the importance of Greenland as a barrier to the free flow of arctic air into the temperate areas of Europe. The cloud streets indicate the wind blowing from the arctic ice into the warmer Atlantic. The east coast has a belt of loose ice with whirls, showing that it is slowly melting.

▲ 2.5 (24/4/84, 17:08 PM, 2; 4; 3)

COASTAL WHIRLS

There is ice close to the Labrador coast most of the year. In April extensive breakup is under way (because of the warmer undercutting sea) even though the wind is often from the north. In the case illustrated here there is a cyclonic whirl which is not visible because the tenuous cloud in it does not contrast with the bright detail of the snowy ice floes beneath. Cloud streets are intense in the cold northerly wind blowing from the ice to the less cold sea. In the infra-red picture, middle, the whirl stands out against the ice and sea, which are warmer than it is, and the Channel 3 picture, right, shows it more starkly because the ice and snow appear black. Such a whirl would have worried Viking adventurer and Victorian explorer alike.

The Mediterranean Sea nurtured our ancestral culture. To them it was The Ocean. It was full of unpredictable winds, now visible to us in the shallow layer of air held under an inversion and trapped like a lake by the surrounding mountains. Each coast becomes known for its particular eddies, many of which are described in ancient classical literature.

Air pollution from the Po Valley invades the Adriatic Sea; Corsica holds off the cloud and an air jet between the islands of Corsica and Sardinia set up a whirl.

▼ 2.6 (23/1/83, 14:31 PM, 2)

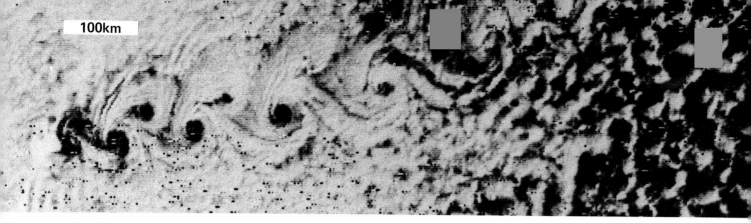

100km

▲ 2.7 (18/9/75, 10:50 AM, VIS)

FOSSIL CLOUDS

Low clouds are often very slow to change form, particularly over the smooth and level sea surface. As they are carried along by the wind they reveal past motion even though it ceased some time back. This is unlike the patterns in streets or a cyclone, which indicate vigorous motion in progress. We call these indications of past events fossil formations. A good example is provided by the vortices in the lee of an isolated island when the air is too stable to flow over the top of it. Although the vortices close behind the island have a substantial rotation, they slow down not long after they have been shed from the island as the next one on the opposite side comes into being. The fourth and fifth have almost ceased to rotate, but the pattern is carried far downstream before it vanishes, 700–800km (435–500 miles) from an island, Jan Mayen, which is only 55km (34 miles) in length (2.7).

Many previously unrecorded configurations have been discovered by our eyes in the sky. We show one below (2.8) which represents the winding up of the boundary between two cloud areas. There is no high cloud in the neighbourhood, and we do not know the mechanism of formation of this three-dimensional looking piece of Nature's artwork. The roll has a radius of 100km (62 miles).

THE ROTATING EARTH

Our planet is unique among those in the solar system in having a partially cloudy atmosphere and a rotation speed which makes possible a very wide range of sizes of rotating cloud systems. Although anticyclones are the largest, the frontal and tropical cyclones are the most powerful and sometimes devastating. Many Atlantic hurricanes draw the polar front into their circulations and move towards Iceland and Europe. This one found no front waiting to renew its energy, and it dies in mid-Atlantic. Like most hurricanes, it had an eye, a clear calm centre. It is seen below (2.9) at 37 deg N, 40 deg W.

In the Indian sector they are usually called cyclones; in the Atlantic they are hurricanes; in the western Pacific they are typhoons; but they are all the same kind of storm. Our second case produced the lowest sea-level pressure ever recorded (except possibly in a tornado, which is a much smaller scale phenomenon with its 'best' cases over land). It reached down to 870mb on 12 October 1979 at 0600 GMT at 17 deg N and 138 deg W. The cloud top was so high that it achieved the extraordinarily low temperature of -93 deg C, with an area about 300km (186 miles) wide showing a temperature below -80 deg C. The picture opposite (2.10) shows it by early morning sunshine a few days later as it attached itself to the polar front and approached Japan. The centre is now at 25 deg N, 128 deg E.

▼ 2.8 (15/6/80, 14:52 PM, 2)

▼ 2.9 (26/11/80, 21:00 PM, 4)   2.10 (18/10/79, 06:00 AM Tokyo, VIS) ▶

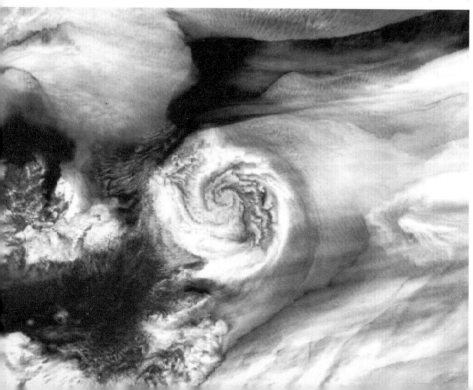

100km

▲ 2.11 (14/12/86, 15:57 PM, 4)

Temperate latitude cyclones are often called depressions or lows and the ones in the North Atlantic inspired the Norwegian frontal theory which described the life history of the stereotype. Our example shows one in the typical position of the Iceland low, which is where lows occur with about the highest frequency. This one produced the lowest sea-level pressure ever recorded in the Atlantic outside tropical storms – about 916mb – and certainly the lowest since satellite observations began to provide pictures.

At this stage it looks cuddly, but it is growing rapidly, sucking polar air into the Atlantic with hundreds of stormy showers behind it. The whitest cloud is the highest, and we see that in the middle it has moved ahead of the low cloud, forming what we call a 'shelf cold front'. In the warm sector to the south there are only low warm clouds with dense warm front clouds moving ahead of them.

Greenland is in the north of the centre, while Iceland is already half covered by the warm front cloud. The cyclone was most severe nine hours later.

Twenty-four hours later it is seen in the next picture. By then the most severe showers in the polar air were close to the northwest of the British Isles and the cold front had cleared the western half of Iceland. This is a composite picture taken from two successive passes 100 minutes apart by the same satellite. The Norwegian coast is visible in the east, and the snow-capped Alps are seen in the southeast.

DUST AND GLINT

Dust can be distinguished from cloud by its appearance but it may be confused with glint – the reflection of sunlight from the sea. When the sea is very calm parts of it are so mirror-like that they reflect a beam so narrow it misses the satellite, and that piece of sea is recorded as very dark. But if that beam does enter the satellite's radiometer the piece of sea appears very bright. Thus, what appear to be the very brightest and very darkest bits of sea (at least to our eyes in space) are not far from each other and are actually very similar. (Of course, this is not the first time man has been fooled by mirrors!) Where the sea is ruffled by wind the beam of sunshine is spread over a wide cone of light and the sea looks the same as if a layer of dust were in the sunshine above it. The incident shown

here was the beginning of a great outpouring of dust from Africa lasting several days, and carrying dust as far north as the sea west of Ireland. This highly polluted air was approaching an exceptionally clean airmass to the north. Its cleanliness was demonstrated by the presence of ship trails, which only appear in very clean air (see Sub-suns and Pillars p 157 and Transforming Sea Fog p 170).

There is glint on the east side of this early morning pass of the satellite, in the Mediterranean, and it is brightest close to the coast of SE France and NE Spain. There the sea is very calm, and there is a dark patch between Majorca and Sardinia, where it is also calm. The dust obscures the coast of Algeria and covers much of the near Atlantic. The ship trails are in the cloud stretching westward from Brittany.

▼ 2.13 (21/8/80, 08:36 AM, 2)

300km

A BROADER VIEW

Satellites can very effectively fill the gap between what can be seen by a single observer and the size of the entities on a weather chart – cyclones, fronts, etc. Radar has made a big step forward in filling this gap in sensitive areas, but it has always been limited by the range of any one instrument. There is also an added dimension that only our eyes in the sky can offer – an overview of our exquisite blue planet.

Meteorological satellites are excellent science, brilliant technology and exciting to use. They have enormously increased our familiarity with the atmosphere, and are gradually transforming meteorology. But perhaps their greatest contribution is in extending our comprehension beyond the visible horizon.

We have to regret that professional duties compel many users of satellite information to seek ways of incorporating it in numerical form into the exciting and powerful computer models on which our advanced forecasting operations have become dependent. There is still a need for skywatching from satellites. The infinite variety is always full of surprises. The sky places a challenge before us not only because 'it is there' but because we are in it, and there is a process of getting to know it which has hardly been taken over by computers because it is a time-consuming process.

3
STARTING POINTS

MOIST – RELATIVELY AND ABSOLUTELY

We have a wonderful mechanism for keeping our body temperature stable: when we get too warm we perspire and the evaporation of sweat keeps us cool. We have another wonderful mechanism for keeping pollution out of our lungs: this works by keeping our bronchial tubes moist. It means that every breath must be saturated with water vapour, and every kg of air we take in must be topped up to contain at least 42g of water vapour before we breathe it out.

Cold Arctic air at -10 deg C contains only 1.8g of vapour, even when it is saturated, and so people in cold climates must provide about 40g of water for every kg of air breathed. In some regions they have evolved noses which keep cold and recover some of that water by condensation just before it is breathed out. A dweller in the tropics, where the temperature may be 35 deg C but the air may already contain 18g even if only 50% saturated, only has to provide 24g of vapour for each kg of air breathed.

This illustrates the very important fact that at only 50% relative humidity in the tropics the air may nevertheless contain about ten times as much water vapour as air with 100% relative humidity, ie saturated, in the Arctic. At higher altitudes the difference is reduced, and at the heights where jet aircraft fly the air is very dry. The traveller will therefore find that more drink is needed on long flights than at home.

Humidity is measured in two different ways. The relative humidity is the percentage (%) of the maximum amount of vapour the air can contain that is actually present. The absolute humidity is the actual amount of vapour present and is expressed as the number of grams of water vapour in a kilogram of air (g/kg): this is called the mixing ratio and the relative humidity is the ratio of the actual to the saturation mixing ratio, expressed as a percentage.

GOING UP AND DOWN

This is about convection. When the air is stably stratified there is no convection: a body of air displaced from its equilibrium level upwards or downwards tends to return. If it is displaced upwards it finds itself cooler and denser than its surroundings and it sinks; if it is displaced downwards it finds itself warmer and lighter.

When the airmass is unstable, the displaced body of air finds itself subject to a force in the same direction as the displacement, which therefore provides a trigger which releases energy from the system. With the slightest push the air goes into convection, and the layers become mixed.

In the atmosphere we have a very important additional possibility: the air is stably stratified as long as no cloud appears, but as soon as any cloud is formed the heat released by condensation makes the displaced body more buoyant and convection begins. In this case the air is said to be potentially unstable. Clouds make possible a whole range of motion patterns which could not occur in a dry atmosphere. The cloud shapes and changes of shape express in visible form the mood of the atmosphere and the basic physical principles help us to read the face of the sky. It's possible to become acquainted with its moods without grasping the physics, and vice versa. To get to know them both together helps us to learn about each.

BASICS

The total amount of cloud in the world remains more or less the same year after year. Therefore the disappearance of cloud must be as important as the production of it. There is one overwhelmingly important cloud-forming process – the ascent of air. At a higher level the pressure is less and so ascending air expands and is cooled in so doing. At a height called the condensation level it has been cooled down to its dew point, and so further ascent causes condensation of water vapour into tiny cloud droplets which form on condensation nuclei of which there is normally a plentiful supply in the air naturally.

When the temperature is so cold that droplets are replaced by ice crystals we call this glaciation or freezing. If air descends and is accordingly warmed the droplets evaporate, and evaporation is complete as the air passes through the condensation level. Droplets do not normally freeze as soon as the temperature in ascending air falls below 0 deg C because there are not normally the right kind of nuclei present in the droplets to initiate freezing. The droplets therefore become supercooled and freezing does not begin until the temperature is down to -15 degrees or less, and then it occurs almost exclusively in the larger droplets. If some droplets freeze and then the air begins to descend and passes through the condensation level, of course the droplets will evaporate: but if the temperature is below 0 deg C the ice crystals do not immediately evaporate because they exert a lower vapour pressure. The air must be brought down to the ice evaporation level before the ice will evaporate.

But if the descending air comes to the 0 degree level before it gets to the cloud base the crystals will begin to melt, and so we call that the melting level. Aviators call it the freezing level because above it the supercooled droplets freeze on impact with the aircraft and accumulate as solid ice on it, adding weight and spoiling the aerodynamic shape.

Thus we have two anomalies: droplets do not immediately freeze on passing up through the melting level (0 deg C), and crystals do not immediately begin to evaporate on descending through the condensation level if the temperature is below 0 deg C.

When droplets grow in size and crystals grow or accumulate into snowflakes or supercooled droplets freeze into a hailstone as it falls and collides with them, they become fallout, ie drizzle, rain, snow, or hail. If they have far to fall they sometimes evaporate in the clear air below cloud without reaching the ground.

Each parcel of air has its own condensation level which is determined by the amount of water vapour in it. The maximum amount possible is very small in cold air, but it is ten times greater in cool temperate regions than in the extreme Arctic, and ten times greater still in the tropics. Clearly the amount of rain, snow, or hail is much greater in warm climates than in cold ones.

Fallout is common, and so the air at high altitudes contains less total water (vapour, liquid and ice crystals) than at low altitudes. Cloudy air does not have to descend to the level of cloud base to be evaporated, and so there are large areas of blue sky in the regions of downcurrents, or subsidence. The skies are therefore only partly clouded and provide us with ever-changing scenes with a gorgeous backcloth of blue.

When air is warmed we know from experience that wet objects become dry more quickly. This is because the actual mixing ratio remains the same while the saturation mixing ratio increases. The relative humidity is therefore decreased and the air feels perceptibly drier.

When the ground is warmed in sunshine and the air is warmed as a consequence, considerable amounts of water are usually evaporated from the ground, either from the vegetation (which keeps cool that way) or from wet areas. Thus convection usually warms the air and increases its water vapour content at the same time. If the supply of vapour is large this may lead to a lowering of the condensation level and the formation of cloud, but more often on sunny days the heating is more important, so although the convection carries moisture upwards the warming actually decreases the relative humidity, and sometimes even evaporates a layer of low cloud which was there before. We must also remember that the air close to the ground is usually wetter (absolutely) than the air above, and so cloud that forms early on when convection begins may soon disappear by evaporation into the air higher up.

The sunshine disperses fog by warming the air; and it may at the same time evaporate dew lying on the ground, which makes the absolute humidity greater.

CONVECTION CLOUDS

Cumulus

Through the blue sky, and through thin cloud, the sun shines and warms the ground. As a consequence the air close to the surface is warmed; it expands and its density is reduced so that it floats upwards through the unwarmed layers above.

It rises in the form of thermals which mix with the air into which they rise. The condensation level is determined by an average of the humidity all the way up. As a consequence of the mixing the clouds condense at very much the same height over a large area and have flat bases.

The mixing can be seen when the thermals contain smoke and they can be seen to get bigger as they rise, just as a chimney plume gets wider. The first clouds formed are cumulus and their rising tops have an ebullient appearance like a cauliflower with a sharply defined outline. Unlike smoke, which is seen to be gradually diluted as it billows into the air, a cumulus cloud appears suddenly, and on mixing is gradually diluted but suddenly disappears as the multitude of droplets evaporate. Even though the amount of water held in the air is only slightly decreased by mixing, the cloud shows a stark contrast at the moment of evaporation.

Between the clouds the air contains less vapour; otherwise it too would be cloudy. Consequently cumulus clouds are being evaporated continuously all around their sides. The cloud grows at the top because the air inside continuously rises, pushing to the outside. In due course the fresh thermals have all risen to the top, there is no more individual cloud to promote growth, and evaporation takes over.

Cumulus carry heat (because of their excess temperature) and moisture upwards. But in the upper part of the cloud the evaporation may use up more heat than is carried up and the air is actually cooled, although at some later time the heat used to evaporate the cloud may reappear if new cloud is formed in the air by ascent to a higher level.

As clouds grow higher the tops become colder until at some point the droplets freeze and become ice crystals. We call this glaciation, which is the same as freezing, but it does not usually occur until the temperature is well below 0 deg C. The glaciation releases some additional latent heat – about one-sixth of that which is released by condensation. This, however, is not the main influence which glaciation has on the cloud. Ice evaporates less readily than liquid water droplets, and so the ice crystals grow while the water droplets evaporate if they are present in the air together. Thus the ice particles become larger and fallout is more likely. Also, the air may be saturated for ice and not for water, so that on becoming glaciated the cloud may cease to evaporate, which makes further growth possible; and this could not occur without glaciation.

Anvils and the Tropopause

If there is a stable layer (inversion) at some level which the thermals are not warm enough to penetrate, they become flattened out and spread horizontally as at a ceiling. The shape of the cloud looks like an anvil and is so called.

When cumulus are restricted to a layer by an inversion they become flattened into a layer called stratocumulus. If the spreading layer has a clearly identified base well above the base of the original cumulus we call it anvil stratocumulus.

This leads us to ask 'how high can the highest cumulus ascend?' There must be a limit because the temperature decreases with height in the region in which convection cloud grows, and the lapse rate increases from 4 deg C per km (0.62 miles) at typical cloud base to 9 deg C per km (0.62 miles) at 300mb (or 10,000m, 32,810ft), and obviously it could not, even theoretically, go below the absolute zero of temperature (-273 deg C), which would be reached at about 35km (115,500ft) above the ground.

The air would achieve a radiative equilibrium temperature of about -50 deg C as a result of being warmed below by the Earth and losing heat to outer space. This temperature is more or less the same over the depth of several km from around 12km (39,600ft) up to 30km (99,000ft), and the air between these levels is therefore very stable, the lapse rate being about zero. This layer of air is called the stratosphere because it is stratified in stable layers.

The air below is called the troposphere, which means the mixed region, and the mixing is the result of all the convection and other storm clouds. The boundary between this and the stratosphere is called the tropopause.

Usually anvils have begun to freeze well below the tropopause, so that they have a silky appearance, and are described as ice anvils by contrast with stratocumulus anvils which are not usually frozen.

The tropopause is between 14 and 18km (8.7 and 11.2 miles) above sea level in the tropics, and the biggest clouds may penetrate as much as 3 or 4km (1.9 or 2.5 miles) above it occasionally. Mostly the penetration is nearer to 1km (0.62 miles). In middle latitudes there are places where the tropopause takes a step down; this is at fronts where the air is of different geographical origin on the two sides. The tropopause often becomes folded over at fronts and a double or multiple tropopause is produced. This is also where most of the exchange of air with the stratosphere takes place, and it may be a day or two before the old radiation balance is restored; meanwhile there may be two or three inversions at the bottom of the stratosphere.

In picture 2.12 the cumulus lying northwest of the British Isles have large anvils which are carried by the thermal wind (see p 29) towards the north.

Because of the high stability flight in the stratosphere is much smoother than in the turbulent troposphere. The clear air between cumulus is also stable, and much smoother to fly in than the clouds or below cloud base where the dry thermals make flight very bumpy sometimes. (But see also, Billows p 31.)

When cumulus produce rain they become, in a theoretical way, cumulonimbus, which literally means raining cumulus. But the name is usually reserved for clouds organised to give long-lasting showers, and this usually means having glaciated anvils. The different kinds of cumulus are described in Chapters 4, 5, and 6.

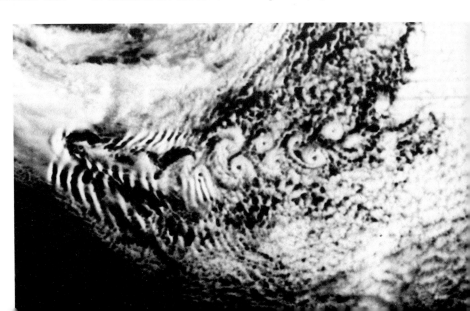

3.1a (23/8/84, 8:53 AM, 2) ▶

3.1a Jan Mayen is undoubtedly the most interesting oceanic island in the world for the variety of patterns it generates in the sky. Others (Madeira, Cheju Do) frequently produce as good vortex streets or (Pico in the Azores, Crozet Island 2.2) wave patterns. But none provide the almost infinite mixture of vortices, waves, trails and dark lines alone and together that we see from Jan Mayen.

100km

▲ 3.1b (17/2/78, 11:44 AM, IR)

3.1b In the centre of this picture is the island of Jan Mayen (71 deg N, 8½ deg W) which is 55km (34 miles) in length. The highest peak, a cone at the northeast end, is 2,277m (7,471ft) and that in the southwest is 769m (2,523ft) high. The neck in between is around 600m (1,969ft) high. The airflow is from the north (top left corner) and is cold, shallow, and blocked by the island. This causes a vortex street similar to that illustrated on page 18 (top).

The cloud streets, which are along the wind, are formed in some places over the ice, but they are increased greatly when the air arrives over the warmer sea and the clouds grow down wind, with a spacing increasing from about 3km (1.9 miles) at the ice edge to 20km (12.5 miles) where the sea is warmer.

3.1c The ice is at about its maximum extent, for it only very rarely reaches beyond Jan Mayen, and it can be seen how the wind drags fragments of ice off the edge to be melted over the open sea.

3.1 (b) and (c) are three days apart and so the speed of the ice movement can be determined by the displacement of identifiable pieces of ice. Displacements from almost nothing to 50km (31 miles) per day can be seen.

Across the bottom of this picture lines of cirrus originating over the high mountains of SE Greenland are being carried from the northwest.

▲ 3.1c (20/2/78, 11:27 AM, IR)

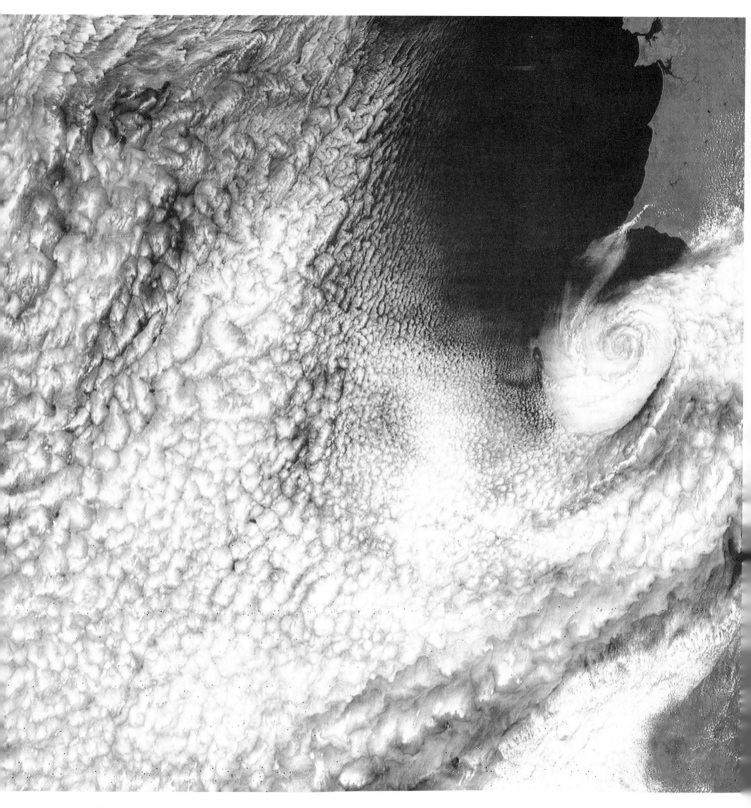

▲ 3.2 (7/6/86, 12:01 PM, CZ5)

3.2 The coast of Portugal is in the northeast and Morocco in the southeast corner of this picture. The wind is from the northeast and falls to calm in the south. The dry air is moistened by the sea and when cloud is first formed it is as very small cells which grow with time to a maximum of around 100km (62 miles).

The mountains near Cape St Vincent separate the flow from the north from that from the east with the formation of a cyclonic eddy. Such eddies are often found in this corner of the ocean between Portugal and Africa.

In this picture the cells are of a type called *closed* with cloud filling the cells. They are more picturesquely often described as pancakes. Picture 3.5 shows some *open* cells.

28

Cells and Streets

If convection is not very strong, as is often the case over the sea where the surface temperature does not get as high as over the land and changes more slowly, and at the same time it is flattened out at a stable layer, the stratocumulus forms cells of a fairly uniform size over a large area. As time proceeds the cells often grow in width, so that having started with a width of 1 or 2km (0.62 or 1.25 miles) when the regularity first appeared the predominant size may become as large as 100km (62 miles) without much increase in the depth of the containing layer. Such large cells are characteristic of light winds, as in a col.

Where there is a moderate or strong wind the only cells which last without much change if they are not raining are streets. They are lined up along the wind shear, which is the same as the direction of the wind near the surface. Streets give a very clear indication of the wind direction in satellite pictures.

The width of streets increases downwind but does not achieve the dimension of large cells. The increase in width is associated with the increase in temperature of the surface as when air moves from ice or cold sea to warmer sea, or from the sea inland by day.

Cells assume many different forms. Sometimes they are closed, like pancakes pushed up against one another with little or no space in between. These have the strongest upcurrent in the middle, where the cloud top is a little higher, and sinking motion around the edges. Clouds of this form indicate that the stable layer at the top is a strong one and sharply stops the upward growth of thermals. It also indicates little or no subsidence in the layer, and no fallout from the cloud; otherwise bigger gaps would appear as a result of the warming or loss of liquid from the cloud.

On the edges of very large closed cells there are often indications of cyclonic motion where the air at the top of the sinking region converges into the downcurrents. In these cases the loss of heat by radiation from the top of the layer is just as important as the gain from the warm sea, and we call it cold downward convection by contrast with hot upward convection from a warm surface. In many cases such as sea fog, where the air is moving towards colder sea and is becoming colder rather than warmer, the radiation is all the time cooling the layer at the top and the sea, paradoxically, is warming the air so as to slow the rate of cooling.

Open cells appear mainly when the clouds are producing rain. As soon as the rain falls into clear air it cools it by evaporation into it. This produces a downdraft which is added to by the weight of the rain which drags the air down as soon as it achieves its terminal fall-speed. The downdraft spreads out on the surface and covers it with a layer of cooled air which has to be warmed up by fresh thermals before new cumulus can appear in it. Thus a large part of each cell is cloud free.

FRONTAL CLOUDS

Airmasses

Air with much the same characteristics of stability and humidity over a large area we call an airmass. Cold and cool airmasses are often filled with convection cloud with cellular or other patterns which are much the same over the area. Between airmasses of different quality we often, indeed usually, find a belt of clouds which may be anything from 1km (0.62 miles) to 200–300km (124–186 miles) wide. These are fronts which are the sloping interfaces between airmasses usually with the warmer mass sloping up over the cooler one.

Sometimes new fronts are formed within advancing cold airmasses by the alignment across the wind of showers which seemed previously to be more scattered. Showers seem to propagate themselves better that way, and the arrangement is called a trough (of low pressure). On other occasions the cold air is rapidly subsiding so that the upper cloud is evaporated and all that is left is a rope cloud, so called because the boundary between the airmasses only remains in a shallow layer at the surface and only a thin strip of cloud remains, although the boundary may still exist over hundreds of kilometres.

The Thermal Wind

This section describes a temperature effect which is quite distinct from the thermals in convection. It is common for ice anvils to be formed on big cumulus which are raining, and these evaporate only very slowly. They are therefore carried along with the wind at their upper level. When this differs from the low-level wind which moves the bases of the clouds the tops are extended along the direction of the shear. They lie, relative to the direction of the streets, along the direction of the thermal wind. This is the difference between the winds at the top and bottom of the layer containing the cloud, and it is caused by the air being colder on the left than on the right (Northern Hemisphere).

The thermal wind is strongest at fronts where the change of temperature is greatest. It causes what is described as vertical wind shear, ie a change in horizontal wind velocity as we ascend to higher levels. Close to the ground the shear is due to the drag of the surface obstacles.

Jetstreams

Warm fronts more commonly produce extensive layers of cloud at the highest levels – the opposite of rope clouds. The stereotype consists of a highest layer of cirrostratus which gradually thickens as the front approaches to become altostratus and eventually nimbostratus as it begins to rain. Because warm air is advancing and the thermal wind blows with warm air on the right, if we stand and face the line of the advancing front the thermal wind blows from our right to our left. Likewise at a cold front the thermal wind comes from right to left if we face the front after it has passed over us.

The strong winds which blow along a front at high level are called jetstreams because they are tubes of air moving faster than the surrounding air, above, below, and on both sides.

Cyclones appear close to the tip of waves on fronts and the wave surrounds the warm sector. The wave is unstable and as it grows the cold front closes up on the warm front ahead of it and lifts the warm air to form the broadest cloud sheet of the system on the warm front. Where the cold front catches up the warm front all the warm air is lifted off the surface: it is said to be occluded and the front that remains is called an occlusion. We show examples of fronts in Visibility Problems, p 131, and they are well illustrated in 3.5a and b, and 3.9.

The very brief description just given can be misleading if we are not aware that Nature generates enormous variety, and almost never the same thing twice. Fronts are not a genetic species which reproduce their young; they are a type of motion of which the definition is very blurred at the edges. At training school one of us was told 'You can have any type of weather with any type of front!'

ALTO CLOUDS

The prefix alto- is used to indicate that the cloud was not formed by ascent directly from the surface, but was caused by motion much higher up and has a condensation level far above the lowest cloud base. The original (official) definition gives the name to middle level clouds which are higher than cumulus and stratocumulus, but not high enough to be called cirrus, and not frozen. This became defined as all clouds between two specified altitudes, but the appropriate limits are not the same in all climates. There are several kinds to be described.

Altocumulus

There is no clear visual or other distinction between anvil stratocumulus and altocumulus which may have originated in anvils higher up or appeared out of blue sky, by lifting near a front or elsewhere. Most of the moist layers in which it appears have been formed in the first place as anvils, or by the shearing over of a moist tower formed by very large cumulus a day or two before.

Altocumulus is very much influenced by the long wave (infra-red) radiation coming from and to its upper and lower surfaces. The cooling at the top produces downward convection which gives the layer a cellular structure. These layers appear and disappear over a period of hours or even days, so that some of the cellular patterns are fossil patterns of humidity from the previous appearance. The liquid droplets emit in a whole range of wavelengths whereas the clear air only emits in the same wavebands as it absorbs, and in those heat can only be lost or gained very slowly. Thus the cloud top emits in wavelengths to which the clear air is transparent and is cooled by loss to space. This creates a strong inversion at the cloud top.

Some layers are so thin that the warmth of sunshine can be felt coming through, and this warms the ground, though a little more slowly than if the cloud were not there. The ground then emits more strongly and this warms the cloud base: on a warm summer day the water becomes warmed in this way and is evaporated in an hour or two.

3.3 Condensation is taking place in a thin layer of cloud, and the latent heat released thereby has caused instability. Little puffs of cumulus are rising out of the layer giving towers which are tall compared with their width. These *turret* clouds are called *castellatus*. Those on the right are mixing into very dry air and quickly evaporating: they are called *floccus*. In the distance is a line of castellatus growing out of a very thin horizontal strip.

▼ 3.3 (Wimbledon, 1/8/60, 07:30 AM, W)

Lenticular Clouds and Pileus

These are discussed extensively in Chapter 9 and in Pileus: The Elusive 'Cap' Cloud on p 51 respectively. They have been called altocumulus, but they are in no sense cumulus, but have a very smooth outline. Sometimes the air in which they are formed is stable when clear, but unstable when cloudy: they then sprout cumulus, which we now describe.

Castellatus and Floccus

Cumulus clouds which grow in air that, although stable when clear, becomes unstable once condensation of cloud begins, have a taller, thinner shape than ordinary cumulus, which come through the condensation level with a significant size, excess temperature, and upward velocity. These turret clouds are called castellatus. They most commonly sprout from clouds formed where the air is lifted above quite modest hills, or where it is lifted over a valley or narrow sea in the evening when the cumulus over surrounding areas evaporates, causing the air to sink because it is cooled.

Convection over the sea is often rather feeble below cloud base because no large temperature excesses occur like those over land in sunshine. Small cumulus over the sea is often like castellatus in shape because nearly all the buoyancy comes from the condensation. When the air into which the castellatus ascends is very dry the evaporation of the towers is rapid and they quickly look insubstantial. They are often called floccus because of the similarity of their appearance to a fleece or sheepskin, by contrast with the cauliflower-shaped ordinary cumulus.

Billows

By contrast with castellatus, billows are formed in layers that are more stable than the average, but which are upset by having large shear generated in them. The shear is created when the layer is tilted, as when air flows over a mountain. Paradoxically, the greater the original stability the more rapidly is shear generated to destabilise it when it is tilted. The instability takes the form often described as breaking waves. We call them billows (plural) because they are always formed in a group. If the disturbance is large enough they may produce cloud out of clear air. Most often the upslope motion produces some condensation anyway, and the cloud that is formed is converted into a set of billows which can be seen to go through the growth, and sometimes the overturning process.

Billows are often formed without any cloud being produced at all, and they are the cause of what is called clear air turbulence (CAT). They occur in the stable layers at the tropopause, or at multiple tropopauses if they exist, or when the streamlines are tilted by flow over hills. Of course a tilt downwards is just as effective as a tilt upwards, but then no cloud is created.

In the wave produced over a mountain the shear is a maximum at the crest or trough of a wave, but the instability created may not last long enough for it to develop very far. The billows are formed and then the shear is cancelled out by the return of the air to its original level, and the cloud pattern is fossilised. Many

3.4 Billows are formed when a very stable layer is tilted and shear is developed with the upper layer sliding up over the lower layer. In this diagram we show the growth advancing to the overturning, or breaking, stage of these unstable waves (see pictures 3.13 and 3.25). The shear growth which caused the instability is reversed on the down side of the wave, and it is quite common for the breaking stage not to be reached before growth ceases.

different shapes and sizes appear in the same sky, and their fascination is enhanced by their rapid appearance and change, followed by quiescence.

Billows go through a definite sequence of shapes, although as often as not the development is arrested when only partly completed. It ends with rolls of cloud looking very like the cells of altocumulus. This is as might be expected when we note that the instability is the overcoming of the stable layer by large shear: for they may be caused by a decrease of the original stability, just as by an increase in the shear. Thus if there is already some shear and the stability is altered by radiation mechanisms (discussed for altocumulus) a new set of billows can arise.

The billows are always across the shear. This is different from streets which are along the shear: but streets are the result of the heat source being a rigid heated boundary, and the rolls rotate in alternate directions. In billows the upper and lower boundaries of the flow are distorted so that adjacent rolls do not abut, and all rotate in the direction imposed by the original shear. They are like rollers between the upper and lower layers.

Altostratus

This name is given to a formless layer of cloud covering all or a large part of the sky. The position of the sun is usually obvious although the cloud is thick enough for the edge of the sun's disc to be unclear. It is likely that such cloud will glaciate, either because the size of droplets increases within it or because some crystals have fallen into it from ice cloud above. Typically the mechanisms which form the cloud in the first place also produce a layer of ice cloud, ie cirrostratus, above; or the altostratus thickens and eventually produces copious fallout, in which case it has become nimbostratus, which is commonly seen at warm fronts. We would tend to call it cirrostratus in the early stages if a halo appears round the sun. Similar looking cloud is often produced by the extensive growth of cumulonimbus anvils.

▲ 3.5a (12/9/87, 08:57 AM, 2)

▲ 3.5b (12/9/87, 08:57 AM, 2)

3.5a,b The cold front of this Atlantic cyclone has begun to cross Ireland and the west coast is beginning to appear under the clear sky which it brings. (See also 3.8.) As the cold air advances and becomes deeper, convection over the warm autumn sea produces cumulus. These have grown into four lines of showers between Ireland and the cyclone centre. Further south smaller showers are organised into *open cells*: behind each shower is a clear region where cold air has descended and spread out on the sea and it takes a few minutes for new small cumulus to appear.

In the enlargement we see the distinction between the frozen anvils and the new cumuliform growth which has much more sharply defined outlines.

3.5c This picture was taken in the region of the enlargement (3.5b) about half an hour earlier. It shows a typical polar air mass with shower clouds generated over the warm September sea producing frozen anvils.

3.6 Small cumulus represent convection over a warm surface. Because the air has been well mixed the condensation level, represented by the cloud base, is of uniform height. The growing cauliflower-like clouds have sharp top outlines where the rising air emerges from inside the cloud, and is full of millions of small droplets newly formed in the strong updraft.

3.7 Over the sea the upcurrents are much weaker because there are no very hot areas like those over land in bright sunshine. Consequently the droplets remain inside the cloud for much longer, and have time to grow. The surface tension in the droplet surface squeezes the tiny droplets more than the larger ones so that they evaporate, or later on may be captured by collision with those that have grown larger and fall more rapidly.

Consequently, in these fairly small cumulus over the sea many drops grow to fallout size and we have light showers.

3.8 This is a view of a cold front as it passes and brings a clear sky after a heavy shower. It is a line of large cumulonimbus clouds and it brings freshness to the air, which is now cooler, drier and cleaner.

3.9 When tall cumulus reach up to a stable layer and flatten out, often at the tropopause, the frozen anvils are slow to evaporate and are carried away by the upper wind which may be stronger than the wind lower down. This is particularly the case at a cold front where the *thermal wind*, caused by the close proximity of cold and warm air, is strong enough to form a jetstream. In this picture we see the jetstream carrying the anvils northwards as a cold front passes away behind us while we stand facing westwards into the cold airmass. This deepening of the cold air and the occurrence of bigger showers go together with the jetstream which carries off the cloud tops. We can indeed see the cold airmass coming along.

▲ 3.5c (12/9/87, 08:30 AM, N)

▲ 3.6 (10/2/62, 19:10 PM, SW)   ▼ 3.8 (1/8/60, 07:05 AM, S)   ▲ 3.7 (5/6/60, 13:55 PM, NW, 20deg)   ▼ 3.9 (31/8/63, 18:40 PM, NW)

ICE CLOUDS

Ice Fallout

Water droplet clouds are continuously evaporating at their sides (but not often at their top because of the stable layer there). The evaporation cools the air and thereby creates up and down motions which continue to mix the cloud with the unsaturated air next to it. But ice clouds are inhibited from this mechanism because they have to be created in the first place as droplets at water saturation and then frozen. Since crystals exert a lower vapour pressure than droplets they tend to grow, not evaporate, after freezing, and may persist for hours in equilibrium with the surrounding air. They are drawn out by the air motion to look like hair or fibres, and if the air is damp enough they may grow to a size at which fallout becomes evident.

Thus ice clouds are usually instantly recognisable as such, and this quality is a source of great beauty which we hope we have captured both in visual and satellite pictures throughout this book.

The effect of glaciation is seen in detail and in great variety when castellatus clouds begin to freeze. If their temperature is in the range -5 to -20 deg C (roughly) some of the droplets may freeze, and then they grow rapidly in the presence of all the unfrozen ones and fall out of the cloud as a cirrus streak or fallstreak. This has the appearance of fibres although it is really a streak of falling particles. When there is some shear in the air beneath the original turret the streak becomes curled. Cirrus means a curl of hair. The curls may occur in small waves or mamma due to the variations of humidity in the layers through which the streak falls, for the particles may fall faster if they grow bigger in a moist layer, or more slowly if they gradually evaporate in a slightly unsaturated layer.

If the castellatus is very cold, in the range of -30 to -45 deg C, the freezing may be very rapid, so rapid sometimes that we do not see evidence of water droplet cloud at all. It is believed that ice crystals do not form directly from vapour, and that it is always necessary to achieve saturation for water, at which point droplets form. Droplets freeze spontaneously at -40 deg C or colder, and larger droplets freeze without the provision of special ice nuclei at several degrees warmer than -40 deg C.

Newly formed crystals finding themselves in clear air grow quite rapidly and fall fast enough to make streaks.

Castellatus may go through all evolutions between evaporation as floccus, forming a streak without a head. If ice nuclei do exist in the clear air or in water droplets there is no significant pattern of behaviour which seems to require them to exist, and for practical purposes we may forget them.

Fallstreak Holes

Occasionally a layer of altocumulus which is colder than average because it is higher is not frozen at all. Then it becomes frozen in one place because of the passage of an aircraft or (we have to say, from uncertainty) by some other cause, and fallout begins. The freezing spreads within the cloud and the fallstreaks are drawn together into the strongest part of the downdraft, producing a hole in the cloud layer.

Cirrostratus

This name means a layer of fibrous-looking cloud, but it is used to describe any layer that can be seen to be frozen. We can recognize this cloud type when the sun can be seen through it with a surrounding halo. As it thickens into altostratus, typically at the approach of a warm front, the halo fades and the sun's outline loses its sharpness as most, and eventually all, light rays make more than one encounter with ice particles in passing through. However, this effect cannot be taken as a certain indication that a warm front is passing, because it can also be produced by a layer of altostratus appearing beneath the frozen layer, or sometimes when a frozen cumulonimbus anvil is spread out by shear.

Cirrocumulus

When altocumulus is seen at the same level as other clouds which are completely frozen it is called cirrocumulus. The cloud does not have a fibrous appearance. The cells usually appear much smaller than in most altocumulus, and patches of cirrocumulus add beauty and variety to a sky already containing other frozen cloud forms. There can be doubt about whether the cells are actually smaller or just look smaller because they are higher. The variety often includes reticular, or netlike, clouds which are rather transient. They look as if a large area of rather thin layer cloud had become simultaneously frozen and had begun to fall out to form a large number of fallstreak holes, the holes in the net being formed like the clear rim around the hole.

Cirrus

Patches of cirrus may persist for hours with only a very small change in appearance. When the particles are small and in equilibrium with the surrounding vapour in the air they do not grow, and the cloud acquires a diluted appearance as they gradually fall to fill a greater volume of air. In quiet weather when there are only light winds and small shear they may occupy a depth of 1km (0.62 miles) or more.

With the onset of stronger winds and shear the vertical motion also becomes more important. In subsidence the cloud soon disappears. With lifting the increased particle size and fall-speed produce a more fibrous-looking sky.

In the more vigorous case of a developing jetstream not only are these processes intensified, but the cloud becomes arranged on a larger scale into rows along the front, with fallstreaks drawn out at right angles showing the motion up the frontal surface while the thermal wind is strong and along it.

Clouds formed by other mechanisms will be described in Chapter 12.

3.10 We are looking at the 'heel' of Arabia. The land is very warm in the sunshine, and so cumulus are generated. But this also causes a sea breeze to blow, carrying the cooler sea air from the Gulf of Aden inland for a few miles before it is warm enough for the clouds to appear.

In this desert area there are occasional heavy rains which produce wadis, the name for dry flow channels, making the land look like a river delta, or beach at low tide.

3.11 Over the wide oceans the trade winds carry cool polar air towards the equator, warming it and moistening it, but often rather slowly so that the convection is not as violent as in picture 3.1, although the tendency to form streets is obvious. The clouds are continually evaporating as they moisten the air which has been warmed both by the convection and by the sinking of the cold airmass. Later this air will become the moist tropical air which forms the warm, wet, airmasses of the cyclones of temperate latitudes, producing copious rain as a major component of the continual recycling of water by the atmosphere.

This scene is a few hundred miles southwest of Hawaii.

3.12 Here we see the low cloud in the eye of a hurricane of the Caribbean. The eye is surrounded by a great wall of ascending cloud producing torrential rain. But on the inner edge of this wall the cloud is evaporating into the air of the eye which has descended from the stratosphere. The evaporation causes a great cooling so that there is a veritable Niagara of air descending the inner wall of the surrounding cloud. To enter the eye the plane had to travel through these intense up and downcurrents, and we have to thank the courageous team and their meteorologist, Robert Simpson, who obtained this picture. It may be thought of as a closeup of the centre of a storm as shown on p 18 (2.9).

3.13 This closeup of billows overturning was obtained at the top of the Pic du Midi (2,877m (9,439ft) in the Pyrenées) where the air ascending the wind-facing slope slid over the top of the air coming up the lee side full of cloud.

3.14 On the day of this picture there were two typhoons over southern Japan. The aircraft took off from Kyoto, half way between them. This view shows the spreading anvil top of the smaller, western, storm. It shows the cumulus boiling up around the outside with the much higher anvil spreading out at the tropopause.

3.15 At some cold fronts the cold air is subsiding so strongly that no rain occurs. The receding edge of the cirrus in the warm air above the cold air wedge is seen here on a warm July day in Seattle. The rest of the day was sunny and warm in the 'cold' airmass.

3.16 Dust devils are the smallest of rotating convective storms. While the basic principle is simple, they show a great variety of patterns. Some have hollow centres with clean air coming down a tube of dust. Some carry away enormous quantities of material. They are usually short-lived (a few minutes only, or less) and have been seen to rotate in both directions although cyclonic rotation is probably more common.

3.17 Layers of alto-cloud may remain supercooled (below 0 deg C) but unfrozen for many hours if no large droplets are formed. But when an aircraft passes through some droplets are caused to freeze by contact. The freezing then spreads by droplets bursting on freezing (like a burst pipe in winter) and the frozen particles grow to fallout size. The fallstreaks draw the frozen cloud together and the freezing cannot spread further so that a 'fallstreak hole' is produced. In this case we see a mock-sun in the crystals of the frozen fallstreak.

3.18 Cirrus may endure for hours in air that is just saturated for ice, and is very cold (below -40 deg C or so). New cloud can only form at water saturation but freezes immediately at such low temperatures and the crystals grow until equilibrium is reached with the air moisture reduced to the ice saturation value. These cloud patches are stretched and distorted by the air motion, and acquire hair-like curls due to their falling motion. Water droplets cannot achieve the same equilibrium and longevity, and so we know that these hair-like clouds are composed of frozen particles.

3.19 This is another example of cirrus patches, this time at different levels growing into different formations, with fallstreaks and lines in different directions.

3.20 A thin layer of cloud all of which freezes at more or less the same time, becomes unstable. The falling crystals are gathered together into a net-like pattern and drag the air down with them. The compensating upcurrents of clear air make holes where the cloud becomes very thin or even has holes made in it, where the updraft has reached the top. This type of cloud is considered rare because it is usually transformed rapidly, and is often soon evaporated as the crystals fall into the drier air below.

3.21 Looking steeply upwards at the fallstreaks from very cold castellatus clouds, we see the fibrous appearance complicated by *mamma* where they have entered a drier layer and the evaporation causes cooling and sinking. The small bulbous fragments at the base are the result.

3.22 Layers of cloud cool by radiation into space and this produces downward convection from the top which is analogous to upward convection from a hot surface. Occasionally the cellular pattern has points of more vigorous convergence at the top of downcurrents, and then any existing rotation is intensified. These vortices can often be large enough to be seen by satellite, but they are mostly too small to be detected.

3.23 Here is the same kind of cloud as in 3.21 but looking down across the arid part of Nevada. There is some low cumulus on the right.

3.24 Over the eastern Mediterranean we often have cumulus over the islands on a fine summer's day with a veil of varied cirrus – white patches in a glorious blue sky. But when seen from above the cirrus looks dark purple, and it is so transparent that the cumulus can be seen very clearly through it. The cirrus is just beneath the tropopause and the aircraft is flying just inside the clear air of the stratosphere.

3.25 From above we often see billows in the top of cloud layers. This patch, captured by Capt Milner over Saskatchewan, is probably formed in a wave over rising ground. In one place the growth has reached the breaking stage, but this is stopped before overturning is complete and the crest of the wave peels off like hair in the wind.

▲ 3.10 (10/2/62, 13:45 PM, NE)

▲ 3.11 (28/8/65, 07:40 AM, SE, 5deg)

▲ 3.12 (1/9/65, 10:00 AM, SW)　　▼ 3.14 (25/9/66, 10:30 AM, E)　▲ 3.13 (3/7/63, 10:30 AM, W)　　　　▼ 3.16 (E. Africa, SW)

▼ 3.15 (12/7/67, 21:00 PM, N)

▼ 3.17 (Washington, 30/11/87, 14:30 PM, SW)

▲ 3.18 (California, N)

▲ 3.19 (25/9/57, 17:50 PM, S)

▲ 3.20 (Wimbledon, 23/5/62, 09:10 AM, up)

▲ 3.21 (W. Wales, 30/7/57, 19:50 PM, up)

▲ 3.22 (Arctic, 15/9/66, 12:07 PM, down)

▲ 3.23 (Nevada, 23/6/62, 08:55, N)

▼ 3.24 (Aegean Sea, 7/12/63, 13:35 PM, N)

▼ 3.25 (Lake La Ronge, Canada, 12/10/67, 18:43 PM, N)

4
CUMULUS

ORIGINS

A bright blue canopy burns through the morning haze. The coolness warms, the air is stirred, a breeze begins to blow. Bits and pieces come and go, form and fade, then form again. First a few, then more and more until the sky is filled with them. They move and change and grow . . . from small to large, soft to hard, puffs to towers. The shapes abound, our thoughts run free in the endless transformation of white on blue. They are the familiar sights, our friendly companions on a typical summer's day.

Cumulus clouds form in the lowest part of the atmosphere, typically around a kilometre (half a mile) above the ground. Up to this level, air motion is governed by the prevailing pressure circulation so the movement of the clouds will closely resemble wind conditions on the ground. Small cumulus may grow several hundred to a thousand metres deep but large, towering cumulus can extend up to 4–6km (2.5–3.75 miles) in height before beginning to lose their normal features.

In most cases cumulus form during the day, evaporate by evening and are absent during the night. The daily cycle of heating and cooling produces these clouds, and they are a small-scale example of the driving energy for all weather – heat. The sun is its source and the atmosphere serves as a giant transportation system to redistribute the local surpluses. An interesting exception to the usual diurnal cycle occurs when a constant heat source warms the bottom of a cold airmass, as happens in winter over the oceans.

Heat
How does the heat get from the ground to the air? The sun's radiation is absorbed by the ground, which reradiates it in a longer wavelength that is readily and mainly absorbed by water vapour. In humid air this means rapid heating of air very close to the ground (10–20m, 11–22yd). This effect, seen in reverse, occurs at night when a shallow layer of radiation fog forms by the cooling of air near the ground. In dry air the heating is spread over the whole lower layer up to 500m (550yd) deep because the radiation has to travel much farther before being absorbed by the scarce water vapour molecules. Once the air has warmed to the point where it becomes unstable and is thus able to rise by its own buoyancy, individual parcels of air known as 'thermals' begin to rise. This process of convection is the primary method whereby the atmosphere transports heat upward, even when there are no clouds present.

Thermals
Convection begins with rising parcels of warm, buoyant air originating near the ground. When they pass through the condensation level they form small cumulus clouds with flat bases and rounded tops. The top is sharply outlined as the rising dome continues to form new cloud droplets. At the same time, small bits of cloud can be seen below or around the top, indicating continuous evaporation around the cloud's periphery. The evaporative cooling causes a sinking motion which can be seen by watching individual nodules form at the top, rotate outward, then sink around the rim. Each mixed part settles down to its own equilibrium level while the warmer core continues to rise. The end result is a trail of leftover cloud material that merely drifts with the horizontal flow. The top continues to rise until evaporation finally overtakes condensation or until it spreads out at a stable layer.

A cumulus sky is a constantly transforming vista where change predominates over movement. If you glance casually at such a sky you may be tempted to presume that things are more or less the same as before, but in fact, everything is changed, every single cloud is different! The process of formation and evaporation causes individual clouds to change constantly and rapidly. Each cloud may last only ten minutes or so but the sky as a whole will continue to look much the same, changing slowly over hours.

Each cumulus cloud usually consists of several separate thermals rising in succession. Once the air has cooled and subsided a little this site becomes less favourable for continued cloud formation and the remaining cloud material evaporates. However, this patch of air may be a preferred location for later thermals since the air has been moistened by the earlier evaporation.

◀ 3.26 (7/2/83, 14:50 PM, 4) There are many sayings about the North Wind in folklore. In most parts of the northern hemisphere it means a cold wind, particularly in the early months of the year. This scene is the result of a burst of north wind into the extreme west of Europe, almost as if it were a continuation southwards of the scene in 3.1b

The hills of Ireland make their impression on this shallow cold mass which was not deep enough to cross the mountains of Norway. Britain and France are cloudy, Biscay is full of showers, and these extend right down to the Canary Islands. But while the mountains of N Spain thicken the cloud they protect the country beyond where we see wave clouds and streets in sunny broken skies. The very south has clear blue skies.

Catalonia in NE Spain has stratus close to the coast, and as the air reaches Algeria and Morocco new clouds with beautiful wave patterns are carried into the interior.

Over the Atlantic open cells, indicating showers, are prevalent.

4.1 (Utah, 18/8/83, 5:49 PM MDT, NW, 52deg) ▶
4.1 (pp 40–1) This is a sky full of cumulus of all sizes; larger ones are growing preferentially over higher terrain.

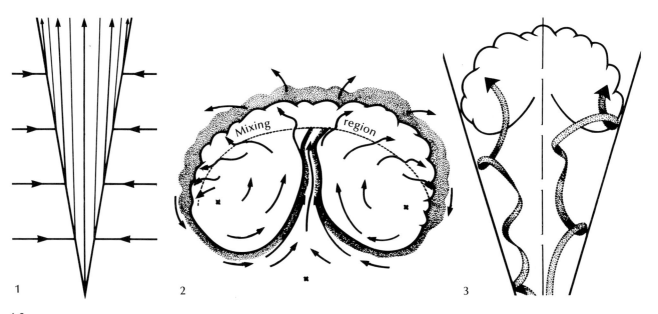

1   2   3

4.2

1. The left diagram shows the average streamlines in a buoyant plume of hot air from a steady source of heat. Individual particles go on very complicated paths and the net effect is to entrain, or mix in, air from all around and the plume is widened.

2. An individual thermal (mass of rising hot air) generates a circulation which drags exterior air up into the middle which is then mixed with air above in the mixing region.

3. Meanwhile the whole thermal is getting bigger and it expands up a cone which is wider than the cone of a hot plume.

The paths of individual particles circulating round the thermal are typified by the tracks shown in the right-hand diagram.

When a thermal reaches cloud level it would typically be two or three times as wide as when it left the ground.

▲ 4.3a (3:25 PM, 62deg)

▲ 4.3b (3:45 PM, 52deg)
▼ 4.4 (Kansas, 11/7/84, 12:52CDT, W, 26deg)

(Alberta, 3/8/81, 3:25–3:45 PM MDT, NNW)

4.3a, b A pair of photos taken twenty minutes apart shows that the sky looks more or less the same although most individual elements are different. The air mass is relatively dry and unstable, leading to tall but thin turrets.

4.4 A small fire is sending up a continuous stream of smoky thermals that expand in proportion with height up to a stable layer, where they slow down and spread out. Each thermal briefly pushes up through the condensation level, forming a small cumulus cloud which then quickly evaporates again. The heat released by the condensation enables the cloud to rise just a little farther, and once it has gone a little puff of smoke will remain above the rest. The subsidence from each cloud builds a weak inversion at cloud base which traps most of the smoke below it, gradually filling this area with pollutants.

4.5 The crown of a large cumulus shows many smaller bumps and dips along the clear/cloudy air interface. As the cloud rises, clear air is being incorporated into it by being trapped in little pockets between the rising nodules. The mixture then slides sideways and down, forming the dome-shaped top. Clear air also mixes into the cloud along the edge since the faster flowing air outside the cloud sets up tiny whirls that draw air into the cloud edge. The small nodules have darker edges because they scatter more blue (darker) skylight back to the observer than the hollows between them which are reflecting white light from the brighter adjacent cloud parts. Double pileus is also present.

4.5 (Alberta, 4/8/83, 8:52 PM MDT, E, 12deg) ▶

Air that rises expands like an ever-widening cone. If the air contains smoke, this expanding bubble can be seen growing wider and wider, until the air reaches a stable layer. If clouds did the same thing we would see huge conical cloud masses perched on tiny bases! Instead, cumulus are almost the opposite and this shows that a powerful additional factor is influencing their shape and size: extensive erosion due to evaporation and outward-turning of cloud material. In small cumulus the entrainment of dry air into their wake soon erodes them, but with larger cumulus the mixing at the top accounts for most of the evaporation that occurs. In fact, the updraft within the cloud may be rising twice as fast as the visible cloud top is. How any one cloud will look depends on the rate of ascent and relative dryness of the surrounding air.

EFFECTS OF OTHER ATMOSPHERIC CONDITIONS

Before the thermals begin there are already several characteristics present in the atmosphere which will influence the size and number of cumulus, when they form and how fast they grow, and even their shapes or structural features. Vertical variations in wind and moisture combined with the degree of instability present all guarantee the uniqueness of each day's offerings.

Stability and Instability

To form clouds at all, the layer below cloud base must not be stably stratified; then thermals rise to the condensation level. Its height may vary during the day as average moisture values change. In general, the continuous upward mixing of heat and moisture early in the day lowers the relative humidity throughout the mixing layer and pushes the condensation level higher. The rapid warming of relatively moist air in the morning can lead to soft, low-based cumulus and fractus. Each one represents the localised condensation of moist thermals within a drier overall air mass. These early cumulus often disappear when further mixing raises the average temperature and dilutes the very moist patches. After a few hours of clear sky and extra warming, regular cumulus will reappear but at a higher altitude than before. When the mixing depth is shallow the thermals cannot rise very far before they are stopped by a stable layer. Heat and moisture build up in the absence of clouds until a critical point is reached and stronger cumulus break out suddenly. On the other hand, a deep mixing depth will permit clouds to form earlier, be more numerous, but also decrease earlier as a result of the larger reserve of dry environmental air.

In the typical case of a clear day with little wind, a profile of the atmosphere would show temperature and water vapour content to be relatively high near the surface but dropping off steadily with height. If the rate of temperature decrease, or lapse rate, exceeds the equivalent cooling within a rising thermal, the atmosphere is unstable and clouds will grow larger. There can be many layers of relatively greater or lesser stability and these can be seen by observing the cumulus as they rise. At locally more stable levels the cloud slows and spreads, and may actually stop there until stronger thermals push through later on. Clouds that enter a more unstable layer will accelerate upward forming sharp, blossoming crowns.

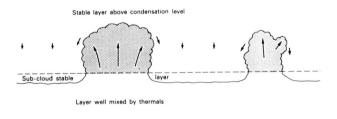

4.6 Displayed are the relative magnitudes of typical upcurrents in small cumulus and the sinking motion in between. If this sinking motion occurs, stable air is carried down below the condensation level, preventing subsequent thermals from reaching it in those places. There is, therefore, a tendency for thermals following their predecessors through holes in the *sub-cloud stable layer* to grow into clouds.

▲ 4.7 (New Mexico, 13/7/84, 12:08 PM MDT, WSW, 23deg)

4.7 Convection is anchored over a mountain here. At more stable layers, cloud material spreads to the right with the winds aloft, forming extensions of the main cloud. The lower one has become altocumulus but the higher one is glaciating as a low anvil with fallout. Continued updrafts are already beginning to penetrate this layer too. The dark cloud fragments seen against the bright tower do not look that way because they are silhouetted. Instead, they are thin (scatter little light) and much closer than the distant tower and scatter only light from the blue sky that surrounds them.

4.10 ▼ (Texas, 27/5/85, 7:13 PM CDT, ESE, 57 deg)

▲ 4.8

▼ 4.9 (Iowa, 1/6/87, 3:53 PM CDT, SW, 57deg)

▲ 4.11a (Nevada, 29/7/84, 6:47 PM, WSW, 43deg)　　　▲ 4.11b (Nevada, 29/7/84, 6:59 PM, W, 52 deg)

4.8 Cumulus castellatus are sprouting in tall towers in the Caribbean. The buoyancy originates almost entirely in the condensation, so that the updrafts are much stronger in the cloud towers than in the thermals between the sea and the cloud base. The still rising tower is sharply outlined but not so the evaporating fragments. In this case the surrounding air is dry enough to evaporate the clouds quickly.

4.9 In dry air, the tops of these cumulus evaporate quickly and become ragged. Only the odd stronger thermal is able to penetrate the stable layer near cloud top, resulting in an accelerated updraft and tall, thin turrets like this one.

4.10 A single cumulus tower has shot up in light winds to form a narrow column of cloud. It has several bulges and dents on the way up corresponding to slight changes in stability with height. The growth was so rapid that evaporation has not had much effect, so the cloud actually exhibits a bit of the conical structure of an uneroded expanding air parcel.

4.11a, b This pair of photos shows an isolated shower falling under a strong cumulus tower. The air, flowing to the right, rises along a single axis, producing a short street of growing cumulus which culminates in the mature shower. In (a) the top is obscured by an altocumulus patch with thin cirrus above, and the cloud gives the false impression of having an anvil top normally seen on the larger cumulonimbus. (b) shows the same system about twelve minutes later. Now, the mid-cloud has moved away revealing a flattening top that is beginning to glaciate as it drifts downwind. To its right is an elongated trail of mostly frozen cloud debris that had been sheared forward from earlier towers whose lower portions have already evaporated. Although the thin clouds above the dark cloud appear high, they are below the cumulus top as seen by the shadow it casts on them.

Moisture

The average moistness of the mixing depth will determine how high the condensation level is and, consequently, whether cumulus form early, later, or not at all.

A dry but unstable air mass can produce high-based cumulus with tall, thin turrets (see 4.3a and b) that shoot up quickly and evaporate minutes later. In some cases the top continues to rise above the already disintegrating lower portions of the cloud.

Moist air masses will not only cause plenty of low-based cumulus and residual layer cloud but also cause the sky to appear more jumbled, more cluttered, with some parts still growing while others evaporate slowly. With greater instability the scene eventually includes larger cloud masses leading to convective showers.

Wind and Wind Shear

When the air is calm near the ground local variations in temperature and moisture remain discrete and can be seen by watching the first cumulus that form. A few may appear over a hill, open fields, or a city but be absent over the cooler lake or forest. Each cloud is a marker of a slightly more favourable pocket of air originating near the ground. Clouds may begin early but will thin out quickly since most of the mixing is vertically and the thermals are soon accompanied by drier downdrafts.

When we toss in a good breeze, things change dramatically. All the local variations are swept away and mixed downstream. Once the thermals become well established, the entire mixing depth is affected and it becomes quite uniform in character. Clouds form later than usual and have trouble growing as their tops shear forward and mix into the turbulent flow. Ragged bits of cumulus fractus appear first, in places where the tops of stretched thermals just graze the condensation level. Gradually the clouds grow, but they are now more uniformly sized and spaced, and larger ones may not form at all since previously moistened pockets of air are mixed out of opportunity's way. It is an interesting paradox that the apparent chaos of a strong flow actually induces the formation of a uniform condensation level and orderly arrangement of cloud elements!

Winds can vary greatly with height, changing in both speed and direction at different altitudes. In the temperate zone the lowest 1–3km (0.62–1.9 miles) are usually controlled by surface pressure systems while above this

▲ 4.12 (Alberta, 1/8/81, 3:00 PM MDT, N, 52deg)

4.12 The sky has a conspicuously bright bank of towering cumulus in the distance. This is the site of very strong new growth which is maintaining a constant supply of dense, bright cloud material. The small, foreground cumulus and flat, stratocumulus patches are all much older and greyer, and parts almost look black where they overlap the brilliant background. The exceptionally good visibility permits cloud details right at the horizon to be visible under the distant baseline. It also enhances contrast usually softened by light scatter from the air (haze). The very dark bases are not, as sometimes presumed, caused by the cloud 'absorbing' the light but are a result of almost all the light entering the cloud top being scattered away in other directions before reaching the base.

an abrupt switch occurs to the prevailing westerlies. An increase in windspeed with height causes all the cumulus to lean or shear forward since their tops move faster than the bases do. Strong shearing can rip the tops off even large cumulus, hurtling the decaying remnants well away from their source. Changes in wind direction can also distort the shapes of taller cumulus, moulding them into lumpy, irregular masses at times. Each cloud is the product of diverse forces whose interplay can be read and understood by the viewer.

NEW AND OLD CLOUDS

The sky is constantly changing and full of variations in colour and pattern. Variations exist between new and old clouds as well as in arrangements of cloud elements. What you see can tell you a lot about the underlying physical properties which have shaped the scene in that particular way.

A cloud that is newly formed or growing continuously contains a large number of small droplets close to the cloud surface which reflect much of the sunlight striking it. This gives cumulus their bright, solid appearance above a shaded base. As portions of the cloud age, smaller droplets evaporate and the remaining lower number of large droplets allows more light through, giving those areas a darker appearance. Such fragments can often be seen around the edges of cumulus or after the cloud has drifted downwind without being replenished by new thermals. This same feature also accounts for the relatively grey or dull appearance of patches of layer cloud such as stratocumulus. Since these clouds are stably stratified, little vertical mixing occurs, and the cloud evaporates primarily around its edges. Internally, smaller droplets also evaporate gradually and the cloud assumes an 'old' appearance.

ARRANGEMENT OF CLOUD ELEMENTS

Cumulus clouds can be arranged or distributed in lines or clusters, streets, or at random. On the large scale, best seen in satellite photographs, this implies regional or synoptic influences such as large land/sea areas or an air mass with uniform properties. The small scale, or a local area that is within view, is highly varied because the interaction of land features, mesoscale factors, and atmospheric conditions causes clouds to be arranged by many different influences. In general, pockets of heat or moisture that always occur are further shaped by lakes, cities, hills, wet ground, etc, then subjected to other atmospheric effects such as convergence or divergence, instability, and the presence of other clouds.

Lines and Clusters
It was shown early on in studies of convection clouds by glider pilots that the updrafts are stronger than the downdrafts between clouds, which would suggest a preference for less cloud than clear sky. In view of this, dense clusters of cumulus, especially large ones, are quite atypical and they suggest a regional influence where the balancing factors to the usual updraft/downdraft relationship may be outside the local view. One example would be a pocket of relatively unstable air which has induced steady lifting there, but is flanked by an area of subsidence that clears the sky again in the distance.

Most lines and clusters result from changes in landform causing preferential heating or lifting there. A ridge will have a line of cumulus on it, a lake may be clear with clouds along the breeze front, open fields may have a congregation of clouds over them while nearby vegetation produces only a few. Patches of moist air from wet ground or previous shower clouds can move downwind and become filled with clouds. Each cluster or line has a definite cause even though that cause may not be obvious or originate nearby.

Streets
Streets are rows of cloud elements lined up roughly with the wind direction. They require a fairly uniform depth of convection over a large area, a common situation over the oceans. Such a bounded layer promotes an orderly arrangement of updrafts and downdrafts which can be more easily controlled by the prevailing flow. This is true even when no clouds are visible (eg fog streets), and there is some evidence to suggest streets form very early, just after the first thermals. This could result from wind drag along the ground which would lean thermals along the flow.

Why should cloud elements form in this manner? The simplest situation is one where a single street forms downwind of a favourable source for convection. The moisture ascends at the source, moves down the flow, as the cloud evaporates, and then provides a preferred environment for subsequent cloud formation. The wind field structure plays an important role in shaping more complex multiple streets. An airstream is rarely completely uniform because the smallest variations in temperature (buoyancy) and topography will cause the flow to split into channels of locally stronger and weaker motion, elongated along the shear.

Clearly defined streets require a fairly dry, usually anticyclonic air mass so that the updrafts, upon reaching the stable layer, do not merely spread as a stratocumulus sheet, but are subjected to continual evaporation. Over land the streets are spaced about two to three times the thickness of the bounded layer and changes in spacing indicate changes in the layer's thickness. While air rises within the streets there is weak subsidence between them which inhibits cloud development and maintains the pattern. In the presence of shear, streets will line up with the shear and can often be seen migrating across the flow at a small angle (10–20 degrees). The sideways displacement indicates either a bias in the formation and evaporation of cloud material forced by horizontal shear or a shifting of the heating below the streets. It might be interesting to speculate about the effect of the sun's angle on street propagation. If the sun shines continually on the ground under a street (a situation that would occur with the sun at right angles to the flow) it would be preserved. A variety of effects can be imagined when you combine cloud height, mixing depth, time of day and year and the latitude of the event.

The street pattern inevitably breaks down with changes in flow strength, uneven heating, or shifts in other atmospheric conditions. Commonly, a day that begins with random clouds, organises into streets that later break down as stronger convection lifts or penetrates the stable upper boundary. As with so many other features of clouds, the beauty of an ordered scene emerges only briefly from the unimposing norm. Its presence tickles our curiosity and rewards our selfless attention and watchfulness.

▲ 4.13 (Utah, 18/7/84, 12:22 PM MDT, WSW, 52deg)

4.13 A narrow line of tall towers marks strong heating over an exposed land ridge in dry but unstable air.

4.14 Cloud streets are formed in the upcurrent in the middle of cells aligned along the wind, which is indicated by the arrows on the side of the cells.

In the cell cross-sections the air is rotating; up the middle, outwards at the top, down the sides and converging at the bottom. These air particles move along complicated screw-like tracks. The clouds promote mixing which continuously evaporates them as they are formed in the upcurrent, and there the motion is more complicated.

◄ 4.14

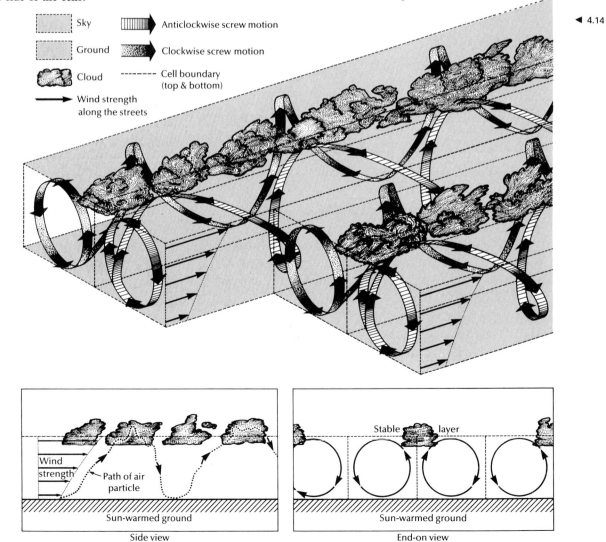

	Sky		Anticlockwise screw motion
	Ground		Clockwise screw motion
	Cloud	-------	Cell boundary (top & bottom)
→	Wind strength along the streets		

Wind strength

Path of air particle

Sun-warmed ground

Side view

Stable layer

Sun-warmed ground

End-on view

4.15 Cumulus are growing along parallel streets. Local variations in topography and winds prevent the streets from being perfectly straight or continuous in all places. As these cumulus grew larger with added heating later on, the street structure vanished. Some much higher, wispy cirrus clouds are visible behind the street, at right.

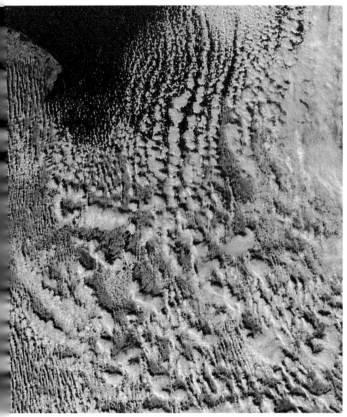

4.16 In a northerly airstream from the Baltic sea over Poland's sun-warmed ground the clouds in the streets are rapidly growing into showers. The streets widen and become fewer.

Where the showers have spread a new layer of cold air on the ground there is either no cloud or new streets of small clouds are growing up close together. Their organisation tells us that they are for a time prevented from higher growth by a stable layer.

In the east these streets have already begun to grow over the warm autumn sea in the very cold air, but on the west side growth is inhibited by the sinking of the air as a warm front approaches.

(Cen England, summer, mid-morning, N)
4.17 Shallow cumulus are forming, then evaporating quickly in a light northeast flow under an anticyclone. The haze top marks the stable layer which is only a small height above the condensation level. The extent and uniformity of the streets results from the wide area of similar terrain below a uniform inversion height.

4.18 Mountains to the left have shaped the flow to their lee. In the cumulus this is evident by the long street, possibly along an axis of greater moisture, while high above it, thin, twisted cirrus streaks hint of a lee wave pattern. A general sinking of air to the lee of mountains would promote street formation and also explain the large clear area devoid of cumulus in the foreground.

A Random Arrangement

The disorderly arrangement of cumulus is most common because so many factors combine and interact to produce them that the sky displays a mix of effects simultaneously. The term 'random' is misleading though, since each cloud is a local phenomenon with an exact cause. What looks like an irregular mix of elements from an observer's glance often exhibits much more organisation when seen regionally, as in a satellite view. A refined variation on random might be a sky full of cumulus with nearly uniform spacing and element size. This can occur if several of the important factors are evenly distributed over a large area, which might occur over oceans or over flat land below a uniform air mass.

LARGE CUMULUS

In small cumulus the thermals are discrete and cause brief, localised variations in cloudiness. As the thermals become stronger they will begin to control the airflow near and below the cloud base so that a continuous inflow can supply the same cloud with a succession of rising pulses. Dry air entrainment is reduced in the lower portions due to the continual replenishment of rising air, and the cloud takes on a solid, dome shape. The evaporation around the outside decreases relatively as the cloud grows larger because the ratio of surface area to volume also decreases and this helps the larger clouds to persist. The steady condensation releases additional heat to maintain the updraft to higher levels and soon the cloud becomes a self-sustaining entity that will not readily disappear as smaller cumulus do. These mature cumulus congestus (towering cumulus) also grow by incorporating the surrounding air (entrainment) and will absorb other cloud patches as they rise through them.

Large cumulus may exhibit several features normally associated with cumulonimbus such as mamma, freezing of the upper portions, and showers or fallstreaks. When dry conditions exist, the fallstreaks may be hard to see but their effect can sometimes be felt at the ground as a sudden cool wind underneath the passing cloud. The shower stage often heralds the end of a large cumulus because the updraft can be quickly overwhelmed by a collapse of cold air dragged down by the rain. If the cloud has an organised inflow that survives this effect, it will most likely continue to grow into a cumulonimbus. When conditions are quite unstable, rapidly growing cumulus may pass through the congestus stage for only ten minutes or so on their way to becoming thunderheads.

When a cumulus tower reaches a stable layer it spreads to form an anvil of altocumulus cloud. Unlike thunderstorm anvils, these patches remain lumpy and unfrozen, and will crumble and evaporate slowly. Another thing that occurs regularly but is seldom seen is a wave-like pulse that travels away from the oscillating cloud top as it comes to rest at the stable layer. A strong pulse can displace the air outside the tower enough to form a small wave cloud in the crest or a clear patch within an existing cloud layer in the trough of this wave.

Warm Rain

Most convective showers over land fall from clouds which are undergoing glaciation in their tops because the rain requires a certain depth and time for its formation, and this is found in cloud towers that have extended up over the freezing level (from 3 to 6km, from 1.9 to 3.75 miles, in the warm season). Very warm air can hold much water and if this condenses in slow but steady convection it produces an abundance of droplets. The smaller ones may evaporate but larger ones will collide to form raindrops, without the cloud necessarily becoming very large. These clouds would show no signs of freezing in their tops and the showers are called 'warm rain'. They are common in the tropics (high water content) and over the oceans (weaker convection) where continuous condensation can occur well below the freezing level.

The same result occurs in cumulus when the updrafts slow down, allowing particles entering the cloud to remain there long enough (30–40 minutes as opposed to the usual 10 minutes) to grow and collide into raindrops. A day could have had many big cumulus without a drop of rain but end in the evening with a quiet shower before finally clearing. It is something of a paradox that a slower-growing cloud may be more likely to generate a shower, the very culmination of the convective process.

4.19 The top of this cumulus dome capped by pileus shows a vertical groove structure in the turrets sometimes seen with very strong upward motion. Compare this with the more common version in 4.5.

4.19 (Nebraska, 1/7/83, 4:20 PM MDT, E, 30deg) ▶

PILEUS: THE ELUSIVE 'CAP' CLOUD

Vigorous cumulus towers push up the air above them, putting a 'dent' into that layer's horizontal flow before penetrating it. If such local lifting occurs in a moist layer, a small pileus cloud may condense in the crest. It is actually a wave cloud formed by the smooth ascent of air within a moist layer which requires only a few feet of lift to reach its condensation level. The air rises and descends through the cloud, which remains in place until the wave crest flattens again, causing the cloud to disappear as abruptly as it had formed. This almost always happens at a stable layer where moisture is higher from the spreading and evaporation of previous cumulus

▲ 4.20 (Arizona, 22/7/84, 5:32 PM MDT, ENE, 52deg)

4.20 One taller cumulus cell is just beginning to flare forward at the top as it becomes a shower. It is almost exactly round (flattened by perspective) and towers about 5km (3 miles) above its base over a distant mountain.

or other cloud patches.

The cumulus slow down as they approach the stable layer and often penetrate it. When they do, they don't simply push through the layer but actually incorporate the surrounding air, including the pileus in the process. The rest of the pileus outside the tower hangs like a skirt around the edge and is either absorbed later or evaporates when the air sinks back to its equilibrium level.

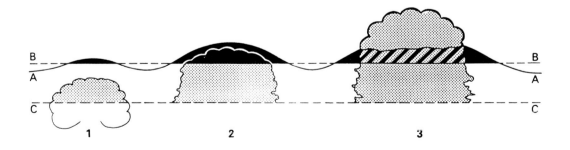

▲ 4.21

4.21 illustrates three stages in pileus growth. Originally there is a stable layer at A and the air just below has a condensation level at B. The thermal has a condensation level at C, and as it approaches A some of the air is lifted above B. The pileus cloud is indicated in black. Occasionally C coincides with B so that the thermal is not visible as cloud.

In stage 2 the cumulus appears to have a very smooth top. If it continues to rise, it will 'eat through' the pileus which becomes absorbed into the cumulus top. The cumulus emerges again (stage 3) often leaving a skirt of the original pileus surrounding it for a minute or two. Any unmixed pileus cloud must ultimately fall back to level A.

▲ 4.22 (Wimbledon, 12/7/59, 20:05 PM, 5)

4.22 This early evening cumulus bank is producing a shower, confirmed by the fallstreaks and presence of a rainbow. The cloud's top has evaporating fragments but no signs of glaciation, verifying that the fallout is warm rain.

4.24 A weak, isolated cell is moving north-northeast after forming over a mountain. The pronounced vertical wind shear has blown its top far ahead of the shower and base, and this anvil plume is now beginning to glaciate. The cell's updraft has several towers (lower right) but these are already weakening as the cloud moves over flatter terrain and never did attain the height of the original surge.

▲ 4.23 (Alberta, 3/7/81, 9:02 PM MDT, N, 6deg)

4.23 A wall of convection to the north impedes a strong southwest flow aloft, forcing air to rise into a mass of pileus with very laminar features. The rising is due more to a blocking than bending of the flow, and this is shown by the wide, nearly flat top of the pileus. The size of the cap also suggests a very moist layer which, in this case, corresponded in height to patches of thick altocumulus in other parts of the sky.

▼ 4.24 (New Mexico, 24/8/83, 7:15 PM MDT, SE, 94deg)

▲ 4.25 (Quebec, 4/4/87, 9:40 AM EST, N, 57deg)

4.25 The process that causes the bending of the flow need not be accompanied by visible cumulus. This is a pileus cap on top of a thermal that has not yet reached its own condensation level. This hilly area northwest of New Brunswick was in a light northerly flow just behind a weak cold front. Moist air had reached the south shore of the St Lawrence River (about 50km, 30 miles, upwind) where fog was present. As this air seeped inland, rose over a group of hills, then settled into this inland valley, it caused a moist layer which formed small, low lenticular wave clouds. The early thermals originated near the ground where the air had been sheltered and was still relatively dry. As this one pushed up against the moist layer it formed a pileus arc draped over the invisible thermal.

▼ 4.26a

▼ 4.26b

4.26a–c (Colorado, 8/6/83, 2:31–2:33 PM MDT, NE, 6deg)

4.26a–c This series shows a cumulus mass rising to meet the pileus cloud. In (a), the cap has just formed with a rather sharp upper surface where the lifting compresses against the stable layer. The air is moving to the right. This pileus is long and thin, suggesting a shallow moist layer that is being bent by several updrafts contributing to the lifting. After 1.5 minutes, the cap in (b) has reformed slightly leftward where new towers are rising to join in. The part just touching at left was the middle tower in the first photo. Notice that although the entire pileus has an overall cap shape, there are also small parts where stronger lift adds to the bending, as seen on the new tower at left. After another 1.5 minutes (c), the leftmost tower has now monopolised the pileus and is trying to penetrate it. The cap now resembles a shroud that will soon be absorbed by the rising cumulus.

▼ 4.26c

▲ 4.27a (Alberta, 28/7/80, 7:00 PM MDT, SW, 30–12 dog)

▲ 4.27b

▼ 4.27c

▼ 4.27d

4.27a–d Powerful early evening cumulus are producing dense, detailed pileus. These four photos were taken to the southwest with the flow from the right, and were about equally spaced over four minutes. The first one (a) shows one strong updraft punching into a moist region to form pileus at two distinct levels. The lower one has already been penetrated leaving a skirt of cloud material which has not yet had time to settle back. The higher one has formed rapidly, causing it to be very dense. The peak of the bending also has two ribbons of cloud formed in thin moist wafers of air separated by drier ones. This laminar structure occurs when the layer is composed of slices of varying moistness, which can happen when the tops of previous cumulus are sheared downwind as moist tongues of air. Each cloud can contribute its moisture at a slightly different level depending on how high its top rises. These discrete moisture variations survive hundreds of miles downwind of convection

that may have occurred the previous day.

The second photo (b) shows three pileus and the tower shearing forward in the strong winds there. Notice that the forward (left) side of the pileus dips down steeply, suggesting additional downward motion in the lee of the tower. The cumulus acts as a barrier to the flow and can sometimes even cause a cap to form in the first lee crest. Between the three distinct layers in (c) we now see several towers rising then leaning forward in succession. Although they rise into and briefly penetrate the pileus individually, they also act as a group to maintain a persistent cap over a nearly continuous updraft. The last photo (d) shows a slight weakening and settling of the cap and underlying cloud. The less obvious laminar structure reflects a smoothing out of the vertical moisture profile, possibly from a general increase in the moisture within the pileus layer.

Pileus are one of the most interesting clouds to watch. They are usually brief, almost ghostly, the silent manifestations of a curious interplay between vertical and horizontal air motions. The complexity of this interplay is shown by the tremendous changes over very short times. Just as pileus require a moist layer to be seen and can tell us about previous convection, they can hint at the future. They also need a strong updraft, and its relative strength determines the degree of bending or persistence of the pileus. Watching these clouds over large cumulus can reveal reliable indications of probable further development culminating in storms later on.

5
CUMULONIMBUS

A cumulonimbus is the grandfather of all clouds. From afar, it stands as a giant mass in three dimensions, a majestic backdrop for all other clouds. Up close, it is an all-encompassing ominous presence that harbours many surprises and plenty of action.

STRUCTURAL FEATURES

The cumulonimbus or thunderhead is a large vertical cloud with a dark, flat base and a partly frozen top that is often extended downwind aloft as its anvil. At the centre is an organised updraft which has lifted the cloud up to a ceiling at a stable layer, usually the tropopause. It occurs as a consequence of deep potential instability and lifting of warm, humid air in a locally favourable environment. This happens when suitable conditions (heat, moisture, instability) combine with a trigger mechanism such as topographic lift, cooling aloft, surface convergence, or the presence of a front, dryline, etc.

A simple breakdown of the average cumulonimbus would display the core region, the anvil top and the inflow/outflow pattern. The core is an active region of heavy precipitation which is forming in a strong, steady updraft averaging 10–30m/sec (33–100ft/sec). In the mature stage there is also a weaker but equally organised and important downdraft, usually just ahead of the updraft when there is vertical wind shear.

The inflow and outflow pattern varies greatly from one cell to another. Outflow can spread forward far ahead of the core with surface winds blowing roughly in the same direction as the cell motion. In other situations the outflow moves with the storm, arriving abruptly as the core passes, or it can be left in the cloud's wake as a cool breeze flowing away from it. The downdraft, which is air that has been dragged down with falling rain and cooled by evaporation, reaches the ground and spreads out as a cool carpet of outflow. The spreading can extend to several dozen kilometres, especially if enhanced by downslope motion, and it may arrive as a cool wind 'out of the blue' from a direction quite unrelated to other local conditions. Inflow is confined to a smaller region, often near the cell's rear, and is marked by a group or line of large, growing cumulus towers that feed into and merge with the core, replacing it as they mature.

Most cumulonimbus develop a downwind anvil, since the winds at the cloud's top are usually stronger than those at mid-levels carrying the cell forward. Less frequently, the anvil forms above the cell but is seen only as a leftover cloud deck while the core shifts forward in a stronger low-level flow. The anvil spreads both forward and laterally as more and more air arrives at the ceiling, then moves away from the source. It is an exciting part of the cloud to watch because it is undergoing constant change. It spreads out as a high, finely textured veil where water droplets are evaporating to leave the ice crystals behind. This gradually thins and brightens the sheet, and helps carry it even farther from its source. This part of the atmosphere is very cold (-20 to -40 deg C) and quite a hostile place for liquid water to be in. Cooling of the cloud layer, combined with a steady growth of ice crystals due to its decreased vapour pressure transform the anvil quickly to a frozen mantle that carries many permanent signs of its original formation. In some places downwind of a stationary convective source, the anvil may be the only clue to earlier events hundreds of kilometres away, the gentle remnant of a powerful force now spent.

The anvil can tell us about the cell's growth rate as well. Weak updrafts will shear forward as thick, lumpy, water anvils. At temperatures of -10 deg C the water may not freeze for hours whereas at -40 deg C the freezing is almost instantaneous. The latter would be more typical of clouds with strong updrafts and the sudden formation of many similar ice particles causes a smoother, more uniform anvil. Updrafts that encounter sudden shearing well above the freezing level also form uniform anvils with sharply defined features. Both strong shearing and rapid freezing act to regulate the more variable effects of updraft/downdraft motions.

▲ 5.1 (Colorado, 7/6/83, 7:07 PM MDT, E, 74deg)

5.2a (Saskatchewan, 15/3/86, 5:25 PM CST, ENE

5.1 This view to the east
shows a cumulonimbus
moving quickly to the
southeast. Most of the
new development and rising
air is to the right of the
heavy precipitation and seen
as a towering cumulus bank
leading up to a billowing
dome which is just reaching
the mature stage. The centre
area is the fully mature
cell which is still active
(notice the unfrozen,
still-rising top) and is
releasing a mammoth
deluge. The fallout
darkens below a certain
point which may coincide
with the melting level.
Most of the anvil is out of
view ahead of the storm
but part of it can be seen
(left) as a sloping deck left
behind in the storm's wake.

5.2a,b Late afternoon convection is winding down in (a) as the March air chills rapidly. The bright precipitation curtain falls as snow until it reaches the melting level, where it becomes water droplets which are nearly transparent to the sunshine striking them. The irregular transition is probably due to downdrafts inducing uneven warming on the descent.

In (b), the melting level is also visible but now appears bright. With the sun behind the cloud here, the falling snow blocks the light quite effectively until it changes over to raindrops.

5.2b (Grantham, 18/4/63, 18:25 PM, W) ▶

▲ 5.3a (6:09 PM, SSW)

▼ 5.3b (6:29 PM, SSE)

(Texas, 27/5/85, 6:09–7:19 PM CDT, 74deg)

5.3a–c The anvil in (a) had formed quickly over the last half-hour atop a large cumulus cluster to the south. The day had been totally clear and it took all that time to build up the heat to break the inversion. Once this happened, tall thin towers sprang up and it seemed only minutes before they were already approaching the tropopause. The anvil's rough edge is a testament to the variations in that original cluster.

The original anvil had all but dissolved in (b), leaving a ghostly skeleton behind. This did not happen gradually as might be expected but took only a few minutes when the cirriform sheet suddenly vapourised under widespread subsidence. This corresponded with an outbreak of renewed convection in which many elements grew equally along a distinct line of rising air. The line was shifting southeast at a sharp angle to the individual cell motion as new towers were being added on the southwest end.

We moved south and southeast about 50km (31 miles) but gained no ground on the expanding line. In (c), the anvil which formed after the previous photo was already just a thin shadow of cirrus and newer activity had pushed another flange out below it. Lower down, continued convection was sending up more towers to repeat the process again. Each storm was a multicell type with several pulses, then a collapse and rebuilding southwestward as a new storm. The distant right one was a smaller, separate cell that had matured on the line's rear. The atmospheric conditions were uniform enough over time and space to permit very similar structures in successive and unrelated cloud events.

▼ 5.3c (7:19 PM, SSE)

5.4 ▲ (Alberta, 14/6/81, 5:10 PM MDT, WSW, 74deg)

5.4 The view is west of a short line of cells moving southeast. Each cumulonimbus in this tight grouping contributes a distinct wedge of anvil material. The distant one was weaker and older (smaller and mostly frozen anvil) but the one at left sits above an active storm and retains the unfrozen detail of rapid formation. Towers rising from the dark base climb, then tilt backwards and fan out. Each rising pulse adds its own piece to the expanding fan like a series of waves hardening the passage of time into place.

5.5 A sprawling anvil is drifting towards us from the west. Winds are light at all levels so the anvil fans out near the cell rather than spreading very far forward. The atmosphere was also quite dry, seen by the absence of lower clouds or small cumulus, and only allowed this cell to form over higher land about 80km (50 miles) away. This thundershower moved southeast (note new growth only on left side of parent cell) then later evaporated away from its more favourable source.

▲ 5.5 (Alberta, 27/7/81, 3:31 PM MDT, NNW, 62deg)
▼ 5.6 (W Texas, 24/5/87, 8:23 PM CDT, WNW, 110deg)

5.6 It took a 115 degree angle of view to gather in the entire overspreading anvil of a distant cumulonimbus. The view is toward west-northwest just before sunset. The core was a small but vigorous and continually regenerated multicell over 80km (50 miles) away and situated on the southwest tip of the cloud deck. The anvil has a sharp edge on both sides but the left half is newer and thicker because storm growth extends the cloud sheet southwards. Fallout and faint texturing exist in several places, with a few large mamma along the distant right edge. Raindrops falling from this sheet all evaporate in the 6km (3.75 miles) descent through dry air. The storm survived for several more hours but never reached the observer's location. The thirsty fields were only teased by faint rumbles and a deceptive facade. Before the modern era of earnestly defended property, the cattle herds would wander towards these storms, expecting to find luscious grazing awaiting them. Man's attention is too focused on his own possessions to allow him to dream of the wider events.

▲ 5.7a (9:07 PM)

▲ 5.7b (9:11 PM)

5.7c ▼ (9:15 PM)

▼ 5.7d (9:19 PM)

▼ 5.7f (9:28 PM)

▲ 5.7e (9:23 PM)

▼ 5.7g (9:34 PM)

▲ 5.8 (Colorado, 21/6/80, 8:00 PM MDT, ENE, 74deg)

5.8 A few strips of altocumulus are dwarfed by the golden wall of a receding storm complex. The view is east around sunset and the horizontal illumination reaches far under the backhanging anvil to expose the falling rain, both below the base at right and from the anvil, seen as the soft yellow glow above the baseline. Higher up, the anvil is mottled as mamma and pockets of subsidence hang into the sunlight. This area appears to be above the storm but is actually many kilometres closer and merely compressed by perspective.

5.7a–g (Alberta, 9/7/82, 9:07–9:34 PM MDT, ESE, 30deg) The opposite series of seven photos tracks a strong updraft pulse through its life-cycle. The storm dropped golfball-size hail for about an hour but had formed late in the day and died abruptly. The cloud is a multicell type with rather large, long interval pulses and is seen to the east-southeast in the warm light of a setting sun. All photos had a 30 deg angle of view (cloud was about 100km (62 miles) distant) except the last one which was slightly zoomed in.

This potent storm formed at about 8:30 pm as a completely isolated event on a day with no other convection. Growth was extremely rapid, catching the weather office (and me) by surprise. In (a) the first few pulses had formed an anvil with a large, new one coming up at the southwest flank. To its right was a sloping line of growing cumulus feeding into the system.

The powerful new updraft surge is already poking up above the anvil level and still going strong in (b). By (c) the top area is beginning to broaden and flare out as the great volume of air bounces against the stable layer at about 11km (6.8 miles). The overshooting top is maintained for several minutes by the steady upward surge below. Meanwhile, the storm's centre region, the previous cell now at the mature stage continues to remain strong because some overshoot is also visible there. A massive bulge perches prominently at the storm's rear in (d), signifying a definite peaking of severe properties. The burst of activity seems to have weakened the flanking line which is now less obvious at the rear. The straight shadow cutting across the cloud's side is being cast by clouds far to the northwest and is not the baseline, which is about two-thirds closer to the horizon.

(e) shows the dome flattening to assume the same structure as the rest of the anvil. The updraft's momentum carries it past the equilibrium point at the stable layer then sends the air down again under negative buoyancy. This squeezes the rising air out sideways which forms the anvil's wide, flat flange at the back. The thick rim of cloud around the top is an unusual form of anvil spreading and may be due to an abrupt drop in windspeed as much as 2km (1.25 miles) below the tropopause. By (f) further settling has spread the top over many kilometres. Updrafts continue, but at a slower rate. A few larger cumulus are faintly visible at lower right but they do not appear to be feeding as directly into the main cloud. In (g) these towers do not look organised and they later failed to mature significantly. Now the anvil is thinner and beginning to lose detail as it glaciates slowly. The storm is passing through maturity to a period of gradual decline but its spectacular structure and rare beauty live on.

61

Why Every Cumulonimbus is Unique

All clouds are unique in the sense that their details don't ever repeat exactly but the similarities predominate within each cloud type. The cumulonimbus fills the third dimension (and the fourth – time) and this produces an almost unlimited number of possible shapes, sizes, and colours to observe. Moisture affects instability which affects cloud height which comes under different shearing forces at every level, giving us a new result with each set of circumstances. And there are many other variables contributing to the outcome. Changing even one of them alters the recipe and the visible result. The atmosphere is always on the move, mixing and varying ingredients, forming each day's special environment to spur or quell a thermal's ascent.

During the day, conditions are changing also. An early shower with low anvil and awkward build hides the potential of a late-afternoon storm with intense updrafts that punch into the stratosphere. All it needed was a little more heat! Under the influence of the jetstream changes from day to day are greatest, but when conditions aloft stagnate separate cells can look quite similar. This is especially true for widely spaced isolated clouds where no interaction is possible. Looking for similarities between different cumulonimbus over various time periods can be rewarding since the remaining differences become more obvious and can be more easily traced to a particular change in atmospheric conditions.

TYPICAL LIFE CYCLE

An average cumulonimbus cloud begins as a large cumulus tower or cluster of towers that continues to grow and organise into a more or less single mass with a single, larger updraft which usually exceeds 10m/sec (33ft/sec). Occasionally this cloud can form from convection originating at the mid-level (altocumulus castellatus) in air which has become unstable from the advection of cold air aloft. Acceleration of the updraft due to the release of latent heat organises the separate thermals and shapes an inflow pattern that draws in the low-level flow from miles around. This early stage is characterised by the fact that all the air is rising, and raindrops that are forming are held aloft in the updraft.

The mature stage is initiated by the onset of precipitation (which may take 8–15 minutes to reach the ground) and the early effects of glaciation at the top. Much of the forming precipitation is supercooled but it takes a while to transform enough of it to ice crystals to become apparent as a fibrous anvil. Glaciation doesn't begin until temperatures are well below freezing, and happens faster when it's colder or when the droplets are larger. A rapidly growing thunderhead resists the transition to ice because many smaller droplets are filling its rising dome. This also provides a plentiful reservoir of supercooled water for hail to collect. Once larger falling ice particles begin the process, hailstones will grow by 'mopping up' these droplets, freezing them on contact. The size of stones is also affected by the updraft's ability to hold them in suspension for long enough to become larger. An appreciable portion of precipitation reaching the ground in many storms comes from melting hailstones. It is challenging to look for hailshafts, seen as straight, bright streaks within the raincurtain and near the updraft core. The mature stage, with all characteristics at their peak, may last for only ten minutes.

The outburst of precipitation creates a downdraft that either makes the updraft collapse or forces it to shift its position. When this happens, the original cloud enters the dissipation stage. Strong shear aloft will carry the effects of the downdraft out ahead of the cell but with light winds this cool air can arrive abruptly at the surface as a wind surge that spreads out to form a gust front. This is a 'mini cool front', usually with a bank of low, dark, ragged clouds along it. Outflow and rain continue but taper off gradually. Cloud bases become much higher because cloud particles are the most easily evaporated in the adiabatically warmed sinking air, and all signs of further convection are replaced by mid and high cloudiness. However, the storm system may go on to repeat the process with the displaced updraft.

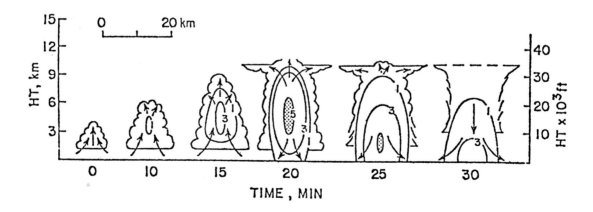

▲ 5.9

5.9 Simplified life cycle of a cumulonimbus cell. Arrows indicate air motion while contours within the cloud show the relative intensity of the precipitation core as would be seen by radar. (*Doswell, 1985*)

▲ 5.10a (1:23 PM, NW, 46deg)

▲ 5.10b (1:34 PM, NW, 46deg)

▲ 5.10c (1:42 PM, NNW, 46deg)

▲ 5.10d (1:46 PM, NNW, 46deg)

CUMULONIMBUS TYPES

The old-fashioned view of cumulonimbus or thunderstorms maintained there were just two basic types, frontal and air mass. More recent research suggests that this oversimplification is too limiting and fails to consider either the role of variable contributing influences or the process of growth and regeneration. The term 'air mass' is particularly misleading since it implies randomness for an event where many factors must come together exactly. The occurrence is not at random, just hard to predict.

One approach would be to group these clouds according to the strength of their updrafts. The updraft is dominant when strong, in balance with downdrafts when sustained, and replaced by outflow when weak. This works well for an analytical approach but is less a 'type' than a 'process' since all three can occur during a storm's evolution or simultaneously in a region.

(W Kansas, 1:23–1:56 PM MDT)

5.10a–e These five photos depict the three main stages of an isolated cumulonimbus seen at a distance of about 80km (50 miles). A weak cool front had slid south into central Kansas the previous evening but had become stationary. The view is in extreme western Kansas looking northwest and the cell formed in light southeast winds and weak surface convergence at the western end of the old frontal boundary.

A small cluster of cumulus congestus had been brewing for about half an hour without much change or organisation. In only minutes (a) the cluster grew together as a single mass and its top began to drift forward and glaciate. The baseline is visible so most of the rain is still in suspension. The inflow enters on the cloud's southwest side and rises as a series of hard towers that are flattening at the tropopause.

This storm is now in the mature stage (b) with heavy rain below and spreading anvil above. The updraft is still active, shown by the crisply detailed dome lifted above the anvil. Many growing cumulus flank the storm, especially on its southeast side where the surface wind is being forced to rise as it nears the cloud wall.

The original cell has entered the dissipating stage (c) and drifts away to the northeast. It is topped by the more uniform right half of the anvil. This same area also has a peculiar fuzzy tuft sticking out above the anvil which formed when a small-scale turret at the top of the updraft shot up and froze instantly. The storm has become a multicell type because new growth on the left end is maintaining the system.

Only four minutes later (d) and the tuft has evaporated into the very dry air above the anvil. The new cell at left is approaching maturity and the momentum of its booming updraft pushes many hundreds of metres into the stable layer.

In (e), the second cell has now begun to weaken and there are few signs of a further regeneration. No new towers are evident on the left side and even the foreground cumulus are thinning out. The anvil has a classic flared shape and extends for about 30km (18.6 miles). This storm rained itself out during the next half-hour leaving only a dense cirriform plume to mark its brief existence.

◄ 5.10e (1:56 PM, NNW, 56deg)

▲ 5.11a (6:36 PM, 30deg)

▼ 5.11d (7:06 PM, 25deg)

▲ 5.11b (6:39 PM, 16deg)   ▲ 5.11c (6:52 PM, 21deg)   ▼ 5.11e (7:22 PM, 51 deg)

(Alberta, 8/8/81, 6:36–7:22 PM MDT, NNE)
5.11a–e The three-dimensional scope of convective clouds enables them to traverse wide ranges of temperature and wind effects to form uniquely sculpted visible tracers of time and change. This series looks north-northeast on a very dry August afternoon in central Alberta. The two showers discussed here upset a prediction for totally clear skies in the midst of an anticyclone and were the only ones to occur that day for several hundred kilometres.

In (a) a sprawling cirrus plume is all that remains of a brief shower whose separate steps are documented in this frozen record. As it drifted slowly southeast to thin out and diffuse gradually, a second cell was being born to take its place. The new cell is sending up several turrets in (b). They are being accelerated upward once they have overcome the low-level inversion. In (c) the first two towers have fully flattened and are already glaciating steadily. The cloud is a multicell type by definition, but is quite weak and has a very long interval between pulses. As the top drifts away, new growth is finally emerging at the back of the cloud. The latest burst (d) is a little stronger and is beginning to curl outward as the last one did. Its bright bulge contrasts with the old and faded remnants of the earlier towers. They are both at about 8km (5 miles) even though the older, closer cirrus looks higher due to perspective.

The overview in (e) now shows both cirrus plumes together. In the cell at right, the cirrus remains are expanding and the cloud is weakening. Evidence of bright, new growth is absent and the base area has narrowed from the rapid evaporation. This newer plume will soon detach as the rest of the lower clouds evaporate. Both exhibit a similar conical structure, one of several hints about their formation. They continued to thin out downwind and completely disappeared several hours later.

All cumulonimbus clouds have a basic cell structure where 'cell' refers to an organised, individual updraft. Subdivisions according to the extent of organisation present are more logical because they correlate well to what is seen, both on radar and in the sky. This approach yields the single cell, multicell, and supercell cumulonimbus types.

Single Cell

A single cell cumulonimbus is one whose life cycle portrays the growth and collapse of a single updraft cell. It may begin as a stronger than average cumulus tower or emerge at the heart of a cluster of equally sized smaller towers that become one larger cell. The lifetime from updraft to a 10km (6.2 miles) height may be half an hour, and rapid subsequent decay has suggested the affectionate nickname 'popcorn cu' for these transient events. They are not very common since favourable conditions for convection usually permit a more continuous stream of thermals, and occur more frequently when instability is shallower or when winds aloft are light. The cell's collapse can initiate new cells along the outflow boundary but they would be separate and independent events.

Multicell

Most cumulonimbus are the multicell type. These differ from a single cell type by having a series of discrete updraft pulses that maintain a more or less steady-state for the cloud's overall strength and structure. The pulses can vary from tens of seconds to tens of minutes apart and can be easily seen and identified as separate, rising towers. Each one may be part of a solid, continuous base but they will mature as separate steps, adding their distinguishable contribution to the storm's precipitation pattern and anvil structure. The cloud system can be quite varied when the strength, interval length, and number of separate pulses are considered, and provides an excellent opportunity to observe the convective life cycle in continuous evolution.

Supercell

The supercell cumulonimbus type is a special cloud system with unusual, almost autonomous characteristics. It will be explored in detail in the next chapter on severe weather.

5.12 The hunk of cloud here is a wide view of a nearby, small thundershower just approaching the mature stage. The convection was slow and disorganised, and new cumulus growth only happened when other showers pushed adjacent warmer air into action.

5.12 (Colorado, 9/6/83, 3:44 PM MDT, ENE, 94deg) ▶

▲ 5.13 (Colorado, 29/6/83, 7:17 PM MDT, SSE, 18deg)

5.13 A large burst of activity has just matured into a well-developed single cell. The view is south with a golden evening light. The cumulus tower at its right is closer and detached from the storm. It may form a separate cell later on. The main cell had formed quite rapidly but such explosive growth can go either way. An equally rapid collapse may occur or the cell may evolve into the more persistent multicell type.

5.14 An isolated storm is made visible by repeated lightning flashes. Most of the light below the base is reflected from falling rain after a couple of bright bolts. The soft glow in the upper portions is where a dozen flashes inside the cloud had lit up parts of the frozen anvil over about half a minute. The anvil's top ends abruptly as clear (dark) sky.

▼ 5.14 (Oklahoma, 2/6/85, 3 AM CDT, NNE, 50deg)

▲ 5.15 (Arizona, 28/7/84, 5:47 PM MDT, W, 74deg)

5.15 A lone cumulonimbus cloud shades the hot sun as it drifts north. The late-afternoon view is west and the air is very dry. The silvery cloud formed as a series of short steps when closely packed towers rose in quick succession and flared forward. The first two, seen as 'points' on the anvil, are beginning to evaporate and only have a few light showers left under them. The third one is still rounded at its top and has moderate showers below it while a fourth one (hidden by foreground cloud) is producing the heaviest rain shaft. The latest one is only about halfway up at the left edge but already has a thin shaft of rain from it. This cloud looks like a single cell in size and strength but is really a compact multicell type with a short pulse interval.

5.16 The hot, dry desert air has finally broken through the inversion here to form an isolated cumulonimbus cell. The cloud is at the mature stage and just beginning to flare and glaciate at its top. The cone-like appearance suggests all rising air with downdrafts well away from the cloud. The updraft displays a columnar structure sometimes seen with vigorous growth (see also 4.19) The convection was triggered by orographic lift over a ridge of mountains in southeast California.

5.16 (California, 16/7/84, 1:17 PDT, N, 23deg) ▶

5.17 Looking east around sunset we see a multicell storm with energetic growth on its south end. The cells at left are passing through maturity, with rain and downdrafts lifting the baseline there. At their right are new towers ready to join the expanding cloud system when they mature. Although cell motion was to the east, new growth was shifting the active core of the system to the southeast. Individual pulses were close together, causing the top to vary almost uniformly from sharp detail to a softer frozen state.

5.18 Surrounded by altocumulus patches, this multicell is being carried by strong winds aloft. Air is entering the system and lifted smoothly along the straight base at right, then rises sharply into a series of well-spaced cells. Where rain begins, the descending air has raised the cloud base, contrasting it with the lower, sharply defined updraft base. This thin, straight part of the cloud lies along an axis of convergence which brings the air together, forcing it to condense in this manner.

5.19 This multicell shows three clear updraft pulses. The first two have formed the usual anvil flanges but the third is a stronger burst which introduces a more sustained, organised phase of the storm's life. It formed on the dryline as the first new cell to the southwest of a squall line.

▲ 5.20 (Texas, 27/5/87, 5:51 PM CDT, N, 80deg)

5.20 shows a multicell cumulonimbus to the north in the early stages of its development. About an hour earlier, another storm had collapsed to the northeast leaving a long flanking line of cumulus towers to fend for themselves. For a long time, nothing much happened. Finally, one tower rose above the rest to become a new, small cumulonimbus, and the cumulus to its southwest began to grow and join in. In the photo, the thundershower at extreme right is trailed by a series of towers. Looking left from the cirriform top at right, each separate pulse is layered in successively sharper, less frozen steps. The pulse interval is short, with four steps clearly visible. The next one is rising at centre as a bulge with a very sharply detailed crown. The dip below its base, often seen under strong updrafts, is the result of air being sucked into the updraft after it was slightly moistened near the base region. Despite the gap on the baseline, this new cell will merge completely with the main cloud mass at right, showing how the overall organisation can shape the destiny of individual parts of the storm system.

WINTER CONVECTION

Cumulonimbus can occur when temperatures are below freezing if there is a nearby source of heat and conditions are unstable. One such situation would be advancing warm air aloft above cold surface air, as might occur ahead of a warm front. In rare cases this can cause sheet lightning in the midst of heavy snow, where the lightning originates in a nearby patch of warmer air. Another situation arises when cold air passes over much warmer water as would occur over mid-latitude oceans or large lakes in late autumn and early winter.

It might be stretching the definition to call the shallower winter versions true cumulonimbus but they nevertheless do exhibit most of the same characteristics as their summer counterparts. They deserve a special look in this section because the unusual combination of arctic air and moist, warm thermals causes some equally unusual cloud formations.

All of the photos presented here were taken near the Great Lakes where extreme contrasts in the air water temperatures are common in the cold season. In dry and stable conditions this leads to a shallow layer of cumulus and stratocumulus over each lake. With increased instability the cloud layer becomes deeper and can extend to 4–5km (2.5–3.1 miles) (where temperatures may range from -20 to -30 deg C) in stronger cells before encountering the inversion.

69

▲ 5.21a (1:42 PM EST, SSE, 30deg)

(Ontario, 18/12/85, 1:42–3:00 PM EST)
5.21a–c shows clouds over Lake Erie with a solid snow streamer in the distance and stratocumulus nearer the shore, where winds passed over less water. The view in (c) is southwest from just inland at the northeast corner of the lake on a December afternoon, with a west wind. The satellite view (b) shows how accurately the clouds map to the water's position, with clouds totally absent over the land between the lakes but appearing shortly after the cold air moves offshore. (a) is the same situation looking south from about 10km (6.2 miles) inland. The patchy low cloud is now out of sight below a solid brick wall of moisture. A continuous supply of warmed air ascends to a very stable layer

at about 3.5km (2.2 miles), maintaining the cloud bank's smooth, unfrozen upper surface. The streamer passes inland left of the view, where it deposited heavy snows before diminishing to a cirriform plume about 50km (31 miles) east of the lake. Such streamers or long, narrow lines of snowshowers form in bands of air rising above the warmer waters. Once established, they can last for many hours, becoming a self-sustaining circulation with surface winds converging along the axis, rising, then diverging aloft. The satellite photo also shows numerous streets in the low clouds downwind of Lake Erie plus smaller ripples across the flow indicating lee waves initiated by the rougher terrain inland and the tightly bounded layer.

▼ 5.21b (Satellite picture 3 PM EST)

▼ 5.21c (2:41 PM EST, SW, 74deg)

(Ontario, 20/10/76, 9:30–10 AM EST, 94deg)

5.22a, b A record cold October morning and light north wind provided the ideal setting for a lake-induced snow streamer. I had awakened at sunrise to see snowflurries falling from a misty blue-white sky over Toronto and was intrigued by its mysterious origin. We headed north up the west side of the cloud street, watching it swell and solidify with every passing minute. After about 40km (25 miles) we turned east and headed straight for the cloud bank. The brown, bare ground of fall was transformed by a magical white carpet into another world. The snows became heavier until visibility approached zero in Newmarket, where mid-morning traffic was crawling around in over 10cm (4in) of snow. On the other side of town the whole process reversed again and soon we found ourselves on bare ground again.

A look back to the northwest through wide-angle eyes in (a) showed a remarkably compact, well-organised line of cumulus rising and merging into a cumulonimbus-like structure. The source was a small lake now about 30km (18.6 miles) upwind and no more than 25km (15.5 miles) in diameter. It was spawning a string of cumulus that drifted south, cooled slowly, then began to drop their moisture in increasing amounts as they approached the Newmarket area. Farther downwind, the snows petered out until practically the entire cloud had been deposited and only the stray flake remained. We headed north a bit, then west through the back of the line and once more entered the brilliant world of winter magic. To the north and northeast in (b) the street's source could be seen almost at the horizon. From it came a line of cumulus in various stages of development. One prominent icy plume had matured earlier from a larger thermal and was now already snowed-out. To its right was a newer cell with a dense,

▲ 5.22a (9:30 AM, NW)

white snow shaft underneath it. One cloud after another glided silently south like an endless procession of vapoury jewels over sleeping thoughts.

These events occurred on 20 October, 1974, one of my earliest experiences with this phenomenon. From a naïve and inexperienced assumption I made the decision to head away from this glorious scene to the 'bigger and better' prospects of Georgian Bay to the northwest. There, I disapointingly found only the impotent presence of a broken stratocumulus sheet. I never saw such a graphic example of the winter lake-effect again and my memories and photos would serve as both a painful and pleasurable reminder of Nature's secret ways.

▼ 5.22b (10:00 AM, NE)

▲ 5.23a (10:03 AM, NNE, 74deg)

▲ 5.23b (10:49 AM, ENE, 57deg)

(Ontario, 18/12/85, 10:03–10:49 AM EST)
5.23a,b These two photos were taken about forty-five minutes apart to the northeast in late morning. A bank of large cumulus was along the south shore of Lake Ontario. When over water in (a), there was enough heat to maintain the clouds in a mostly unfrozen state. Bases were dark (partly due to the dark water surface relative to snow cover) but little precipitation was falling from them despite some vigorous embedded convection. When the clouds come inland in (b) and lose the heat source they cool and condense into snowshowers. Even while much of the cloud is transformed to a wall of snow, there is enough latent heat released to maintain cumuliform tops for some distance inland. The smoky patches in the blue sky of both photos are cirrus-like ice crystal remnants blown downwind from other snowshowers.

▼ 5.24 (Ontario, 9/2/75, 10:40 AM EST, W, 94deg)

5.24 It's a cold, clear February morning. All night long the still air was being filled with bubbling cumulus over Lake Huron just 30km (18.6 miles) away. Then the day breezes began and the clouds rushed inland where they cooled over the snow cover and dropped their moisture in bursts of heavy snow. The squall here resembles a fully frozen cumulonimbus anvil hanging to the ground. As with many snowsqualls, the cell is usually preceded by a cirriform anvil that lowers and thickens just like a regular one as it approaches.

6
SEVERE WEATHER AND THE STORM ENVIRONMENT

STORM TYPES AND ORGANISATION

The storm environment is a very special place in which many different cloud features and air motions combine to produce a more-or-less unified, single entity. It is useful to view this entity not as a cloud or specific thing but rather as an evolving process wherein the various parts are interacting with each other and contributing to a totality. What is seen at any one moment is then a set of visual manifestations of that process, excerpted from an ongoing spectrum of changing features. Although the components are all in a state of flux, the system can be in a balance (steady-state) or in transition. The challenge for the observer is to recognise the interactions, place each isolated or momentary feature in its wider context in space and time, and then begin to sense the whole process through the individual moments of an all-encompassing, collective event.

Convective storms vary widely in size from a single cumulonimbus cell of several kilometres diameter to a highly organised cloud system over a hundred kilometres in extent. They occupy and influence the entire depth of the troposphere, reaching from the lower stratosphere to the Earth's surface. Changes in structure and appearance are often sudden, brief, and rapid but yield some of the most unusual and interesting sky effects imaginable. Since these systems affect a local area over a relatively short period of several hours, they afford us one of the best opportunities to watch and study different atmospheric processes as they compete and strive to balance the forces of nature.

THE INTENSITY OF A STORM SYSTEM

Storm strength is not as easy to assess as it may seem. Defining severity as the storm's impact on us, it includes large hail or hailfalls, wind damage, flooding, etc. On the other hand, intensity might refer more to a storm's structure, rate of growth, and internal characteristics. These features, which would include updraft/downdraft speeds, cloud top height, anvil sharpness, etc, are more easily discernible visually and give the observer the ability to assess a storm's strength from a distance. Although severity and intensity do parallel each other to a large extent, they are not synonymous. Some clouds look fantastic but may produce only weak thundershowers, while other less spectacular clouds manage to inflict heavy damage from

hail and wind. Strength is partly a matter of degree but also depends on the storm developing a life of its own, an independence that becomes apparent in the overall organisation of the system.

The requirements for a system to become severe are quite stringent and only a small percentage of all storms do so. All the necessary factors – a suitable wind and temperature profile in the atmosphere, a large reserve of available energy (the right local ingredients), and a high degree of organisation and efficiency (the right structure) – must be present and work together constructively. In most cases, this means a build-up of favourable conditions over several days, followed by a rapid breakdown of the stability aloft. The longer the build-up without dilution, or the more abrupt a change in stability, the more extreme the conditions and contrasts can become. A storm that then becomes established will feed on this vast reserve and dominate the atmosphere. Many severe events are isolated because the absence of nearby convection preserves the maximum available reserve of heat and moisture for that one system. For the sake of discussion, our analysis is based on storms in west central North America, but many of the same conditions apply elsewhere in the world.

The ideal circumstance has slow-moving high pressure at all levels sliding east to allow the air to warm steadily under a stable inversion for several days. South or southeast surface winds raise moisture levels but drier air remains aloft so that clear skies persist. Winds veer with height from southeast at the surface, to south or southwest at mid-levels, to southwest or west above 5km (3 miles), and are stronger with height, especially above 8km (5 miles). This shearing causes a slanted updraft that carries forming precipitation forward, to fall free of the rising air, eliminating the risk of a premature collapse of the updraft core. The veering winds also help to overlap the cooler, dry air aloft with the warm, moist surface air, creating a more acute interface. Once the ingredients are in place, other triggers can set the system under way. Surface convergence or lifting due to a nearby front or high land area can cause or enhance a storm, pushing it to severe levels. Separate cells may merge into one powerful storm or a cell may change direction abruptly, signalling a sudden increase in intensity. Ordinary storms can also become severe when their environment is favourably enhanced, as when a stream of hotter or moister air enters the updraft. The causes are not always apparent but their effects can usually be identified from the changing cloud features.

◀ 6.1a (4:08 PM, WNW, 30deg)

◀ 6.1b (4:17 PM, NNW, 94deg)

◀ 6.1c (4:49 PM, N, 52deg)

(Alberta, 1/8/82, 4:08–5:56 PM MDT)
6.1a–h Several storm cells that formed in the foothills of the
Rockies had joined up as a north–south broken line that was
moving east across the province. As the line became more
organised, the balance between inflow and outflow focused new
development at its south end where a strong new storm emerged.
This series of eight photos depicts its progress and shows how
various parts function together to maintain a single entity.

In (a), the flanking line at the very south end of the system
has formed a pedestal lowering under an increasing updraft.
All cells to the north are weakening and their outflow enters
the updraft along the wedge-shaped rim. A good surface flow
from the south is preventing the cooler air from advancing
beyond this last cell.

From a little farther south, a wide view of the system
(b) shows that the active area of new growth is really only a
small part of the whole cloud complex, and that rain and layer
clouds occupy most of the line's extent. The pedestal cloud
has lowered further in a second step, forming the equivalent
of a wall cloud for relatively dry conditions. It is so much
lower (and flared outward) from the main base that it appears
unexpectedly bright.

A half hour has passed during which the system has
maintained a steady-state. We have moved east and a bit north
to keep pace with the east-northeast cell motion. Through lower
clouds, two stages are visible in (c): clouds in the background
along the original axis have largely frozen, but new towers ahead
of this are still bright and crisp. The lowered area continues to
provide the most peculiar and entertaining sight of all.

The last ten minutes have seen the new towers in the last
photo blossom into a brilliant crown that ushers in the storm's
severe phase (d). It is now a complete entity with flared anvil,
overshooting top, rain north and west of the lowered updraft
core, and a shelf cloud out ahead of it. The shelf is doubling as
an extension of the updraft base, seen here as a dark, straight
base (lower right) under a bright bank of thick cumulus.

◀ 6.1d (5:00 PM, NNE, 52deg)

6.1e (5:07 PM, NNE, 52deg) ▶

6.1f (5:18 PM, NNE, 52deg) ▶

6.1g (5:33 PM, NE, 52deg) ▶

The storm has a classic appearance in (e). The striated backsheared anvil flange at left is punctured by a large, persistent dome of air that pushes well above the tropopause. There is, however, one small detail that could pose a problem for the future. The wall cloud is elongating and merging with the shelf edge, suggesting that outflow is beginning to advance east, undermining the updraft column.

By (f), the system has aged a bit and begun to transform from a dying phase at left into a new one at right. This propagation was initiated by the outflow pushing east to promote vigorous growth right of the previous core. The old lowering (centre) has vanished and reformed a few kilometres to its east. The bright lower cloud under the anvil at left seems a little out of place; it is lower on its southwest side and thus may be the place where cool outflow moving south meets the surface flow.

(g) The system is now in a second severe phase. New growth on the southeast side is extending the cloud quickly, leaving skeletal anvil remains behind in the northwest quadrant. A steady supply of warm air now enters from the southeast, is lifted at the lowering, rises abruptly, then curves to the right aloft as a parade of points corresponding to the separate updraft pulses. The storm was always a multicell type but its pulse interval has shortened here. A third phase, identified by a slightly separate bank of growth below the main dome, is already underway.

Time has passed and the storm, still going strong in (h), has moved farther away now. The third phase is fully established and although the entire cloud is quite compact and streamlined, it looks a little 'dry' here. It is possible that the system is entraining slightly drier air away from the mountains. The storm is an isolated event detached from the original line. As it slipped away to the horizon, it showed no signs of letting up or losing its majestic structure.

6.1h (5:56 PM, NE, 52deg) ▶

▼ 6.2a (Oklahoma, 21/5/87, 5:59–6:06 PM CDT, NE, 46deg)

▼ 6.2b

This storm formed in a pocket of clear sky behind a group of others to the east and a solid line of thunderstorms to its northwest. The line formed along a cold front moving very slowly southeast into northeast Oklahoma. Cumulus didn't get going until after 4pm but then quickly grew and grouped into clusters of towering elements throughout the sky. Once the line formed, however, all smaller clouds vanished.

In (a), a dominant pulse has exploded upward from within a cluster of weaker cells, initiating the severe phase of this system. The steep slope on the rising column is a good indicator of an extreme growth rate and very short pulse interval usually seen in supercells. Rather than a string of separate thermals, this updraft appears more like a continuous jet.

Only 7min later in (b), the tower has already flattened into a short anvil. Strong growth continues but is beginning to split into two columns that may become a double inflow, another feature of some intense storms. The cloud bank at left is the dense anvil of the line mentioned earlier.

Comparing Severe Storms with the Average Cumulonimbus

Regular thunderstorms respond primarily to instability and do not interact with or receive much influence from their environment. They are simply carried along in the larger-scale flow. By comparison, severe storms often move at markedly different speeds and directions relative to the prevailing mid-level flow. Such special motion is a consequence of the large-scale flow carrying cloud material forward and the system's evolving nature. For example, the updraft core, seen as the tall, boiling, cumuliform dome, often shifts to the right (looking northeast) as air begins to rise more steeply on the storm's inflow flank. The air is actually being circulated through a highly organised structure which can be inferred, at least in part, from the visible cloud features. The same air that had been rising through this core ten or twenty minutes ago is now spreading forward in the anvil or being left behind in low level

outflow around the back. The area where this 'shifting' across the large-scale flow occurs is the outflow-inflow interface, where cooler air is constantly undercutting and displacing warmer air. The continuous updraft maintains steady condensation, but this air is carried forward from its original location and is replaced by a supply of new 'unprocessed' air entering from below. The result is a system whose core may move very slowly even though the anvil is streaming far downwind in a strong environmental flow. In an extreme case, a storm actually can move in the opposite direction if its rapid growth rate causes the core to propagate upwind against the flow. This may account for some of the popular notions about storms 'doubling back' over a surprised observer who thought the system had already passed by.

In severe storms, the large-scale flow does more than just carry clouds forward. Along with setting up the instability requirements, it interacts with the

6.3 Shear normally inhibits the growth of cumulus by tilting them over, and rain generates a downdraft. But these two mechanisms combine into one to generate and maintain a steady strong updraft that is necessary for the production of a severe storm or squall line. Moist surface air entering from the left rises smoothly up an incline and the cloud so formed drops its rain into dry air entering at a higher level from the right. Some of the rain evaporates and cools the dry air causing it to descend under the up-current. These opposed currents exchange their momentum and emerge in the opposite direction. In so doing the storm processes the air mass, exchanging the high and low level air by a 'machine' which moves through the air, deposits rain, and maintains itself.

6.3 ▶

convective circulation to strengthen or prolong the updraft and downdraft. The large-scale flow thus becomes an important dynamical component of the severe storm environment. Strong vertical wind shear is known to be a common and necessary part of intense convection, and indicates the importance of a favourable larger-scale environment. At the same time, many of a system's structural features reflect a high degree of independence within a storm-scale circulation pattern. How these two scales of motion interact to produce the plethora of storm effects is still largely unknown. Fortunately for the sky enthusiast, the visual clues available are an excellent indicator of complex air motions and they can suggest some of the answers as well as valuable questions for further study.

The severe stage of a storm's life exhibits two prominent characteristics that set it clearly apart from more common convection. First, it is usually initiated by a sharp, bursting updraft tower that rises above all previous towers to reach the tropopause in a mere ten or fifteen minutes. This first pulse spreads laterally to form a sharp-edged backsheared anvil but the explosive growth continues and can be seen by the appearance of a nearly persistent overshoot dome above the anvil over the updraft core. At this point, the updraft is no longer a series of rising bubbles, but like a continuous jet which very efficiently transports the warm air upward with minimal dilution. The tremendous volume of air arriving at the stable layer quickly forms a very large anvil that looks much more solid, sharp-edged, and uniform than the fluffy and disorganised anvils of earlier, weaker stages. The so-called backsheared anvil is quite a testament to the storm's power because in a strongly sheared environment this rear flange pushes back against the prevailing flow. The surging updraft, upon meeting the ceiling, has nowhere to go but sideways resulting in upwind motion that can overcome winds exceeding 100kmh (62mph)!

The second characteristic is one that is difficult to describe but has to do with the system's overall appearance of unity. Experienced storm chasers have called this 'getting it together', 'solid', 'tight, compact' and so on. The sky has a special look that isn't seen at any other time. It may be the unusual combination of a massive cloud wall with totally clear sky nearby, a strange mix of disparate elements that, somehow, look right together, or just the proliferation of crisp detail everywhere you turn. It is something easier to feel than to describe or analyse and hints at the power and control the system has.

There are three main types of severe convective events, the multicell, supercell, and squall line, plus a few other special examples. These will be discussed in greater detail in the next four sections.

THE MULTICELL STORM

The severe multicell storm is the most common type and is just a stronger version of a multicell cumulonimbus. Most of the differences are quantitative (stronger updraft speeds, higher tops, larger overall size) but the system is also better organised, streamlined, and persistent. The three stages of cell development are usually all present, with new cells beginning as previous ones mature. Each cell becomes the dominant updraft briefly, resulting in a storm whose core is the current main updraft. In this way, the system is a continuous process with components passing through it as successive towers evolve through their individual life cycles.

A typical multicell storm brings discrete updrafts into the system from the rear flank area (south or southwest side) and can survive for many hours in a nearly steady-state balance between the inflow and outflow. The outflow from behind the system sweeps forward south of the core along the gust front and undercuts the updraft core. The core reforms slightly farther forward (usually just a bit southeast), and the process repeats when the gust front again surges ahead. Due to the pulsing nature of this storm type, the effects at the ground can also be discontinuous. Large hail may fall for a kilometre or two, then stop for a stretch before appearing again. Tornadoes are infrequent and are usually short-lived when they do occur. However, the slightest variations in the system's strength can still be discerned from the constant and rapid changes in cloud features, especially below the base near the updraft. The propagation of the system is often to the right of the main flow and cell motion by an average 20 degrees, causing the system to shift appreciably (and deceivingly) at right angles to its visible orientation along the flow.

▲ 6.4a (6:23 PM CDT, WNW 110deg)

▲ 6.4b (6:32 PM CDT, W 110deg)

(S, 15/5/86, 6:23–8:48 PM CDT)

6.4a–e On 15 May, 1986, an unusual storm formed in
north-central Texas. The day fell between two tornadic days.
A weak east–west surface cold front, which had slipped south
into central Texas overnight, was retreating slowly north ahead
of a strengthening south–southeast low-level flow. The wind
shift line was marked by a few moderate cumulus but clouds
were almost absent on both sides of it. First signs of activity
formed near the dryline as a small cluster of showers about
5 pm. By 6 pm, however, a single, large storm had emerged to
control all further convection in the area that day.

(a) Below a high, uniform anvil are several levels of short
inflow bands where the southeast surface flow is lifted ahead
of an elongated rainwall. New cells are forming along a short
flanking line (lower left) and maturing with sharp fallstreaks as
they move to the right. The multicell character is evident from
the bright slit which separates two distinct cells.

(b) Cumulus extraneous to the system are in the process
of evaporating here as the storm intensifies. The flanking line
has become more solid in appearance; grooves under it in the
lower clouds are aligned with the low-level inflow. The anvil
extends well south of the updraft core due to extensive lateral
spreading aloft and a west–west-northwest flow at 12km (7.5
miles). By comparison, cell motion was from the west-
southwest (4–6km, 2.5–3.75 miles, flow).

(c) The storm has now developed a broad, weakly rotating
updraft as indicated by the curved lowering below the main
base. It is elongated and relatively large, and therefore does
not appear to be related to a typical wall cloud. The system is

78

▲ 6.4c (8:10 PM CDT, NW 110deg)

6.4e (8:48 PM CDT, WNW 80deg) ▶

at its peak strength here, and displays a steady-state, compact structure that resembles a supercell which didn't quite make it.

(d) (page 81) A beautifully detailed scene divides a black, thundering sky to the north from a totally clear, pale blue sky to the south. Air entering the elongated flanking line below the inversion forms a smooth cloud from which towers spring suddenly. Their tops are well below the anvil and cast upward shadows on its mamma surface. The system is weaker now and did not attain an intensity comparable with its grandeur. The problem with the structure seems to originate in the wind shear. Surface winds are too strong (30–50kmh, 18.6–31mph), while mid-level winds are too weak, so the system operates almost like a stationary, small squall-line that is fed by new cells at a point on the southwest end. This point's position varied less than 10km (6.2 miles) over a 3 hour period.

(e) One tower after another rises steeply from the point of convergence at the storm's southwest tip. Miles above, a frozen, flaming spectacle awaits the shades of night as it catches the last rays from a sun already below the horizon.

The system weakened and moved slightly east from 9 to 10 pm, then collapsed abruptly shortly thereafter. A look at the time-lapse satellite film the next morning revealed the remarkable, isolated event. The front of the egg-shaped anvil had expanded and streamed forward while its back edge remained stationary for hours. Then, as if an invisible grip had been released, it broke free around 10 pm and streamed far downwind, evaporating finally the next morning as it crossed Louisiana into the Gulf of Mexico.

79

▲ 6.4f (Satellite picture: GOES visible, 6:00 PM CDT)

6.4f From space, the storm takes on the shape of a giant egg. This is the anvil top which has spread both downwind and laterally. The downwind component represents its persistence over time, while the degree of lateral spreading is the result of both expansion of the updraft region due to propagation, and spreading out of air at the tropopause in the anvil. The overall consistency of shape gives good evidence of the degree of organisation and control of the storm complex over the atmosphere locally.

This satellite image was taken just before the first photo in the series. By looking at photograph 6.4a we can see that the anvil extends out beyond most lower clouds and that the main inflow is concentrated along an axis of developing cumulus, whose point of origin is in the distant left. The satellite view indicates smaller, separate towers lead up to a point (the 'point' at the

cell's lower left edge) where rapid development merges them with the main cloud mass. South of this point, scattered cumulus towers mark the position of the dryline, while far to the southeast other clouds indicate deeper subtropical moisture moving north from the Gulf of Mexico.

The 'ruffles' on the anvil's surface are made visible by the low sun angle. These rather regularly spaced cloud ridges and adjacent shadows result from the discrete updraft pulses that emerge atop the continuous updraft in the storm. Each of these surges leaves a residue of frozen cloud material slightly higher in the atmosphere, and these characteristics persist as the cloud mass drifts downwind. Their regularity could also have been initiated or shaped by the imposition of gravity wave forces along the anvil surface.

6.4d (8:34 PM CDT, NW 110deg) ▶

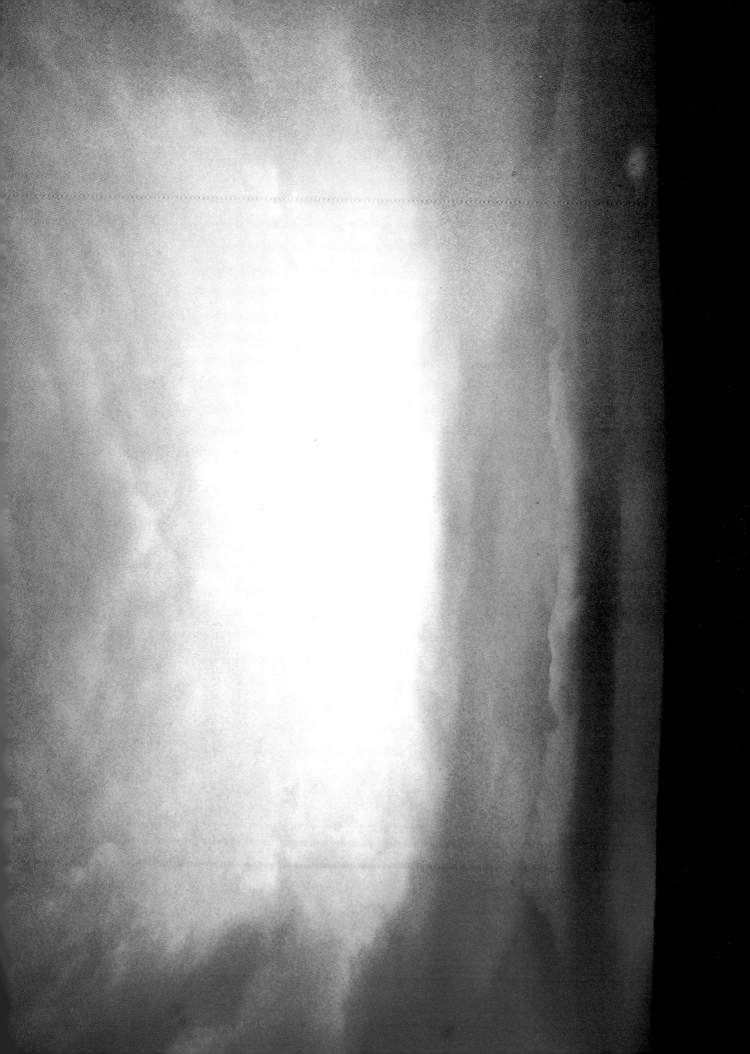

6.5 'Snapshot' model of a severe multicell storm complex, showing cells at various stages of maturity. Dashed lines indicate radar reflectivities (dbz). Precipitation, which forms aloft as each cell approaches maturity (I), becomes widespread at the ground during the cell's dissipation stage (II). (*Doswell, 1985*)

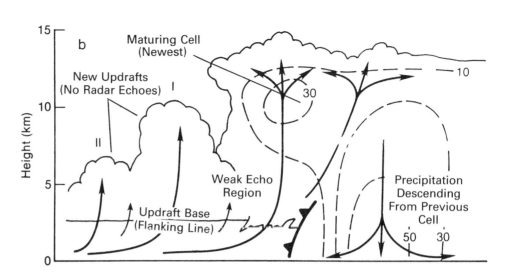

6.6 A boiling multicell storm rises into the golden glow of early evening sunshine. A southeast inflow streams into the complex and rises gradually along the flanking line, (seen as a sloping bank, bottom right, behind lower cloud patches), then rises abruptly at the storm's rear. The system has a large, obvious pulse interval as seen by the three distinct bulges. Upper winds have sheared the first two forward but the rapid development has retained much of their detailed structure. The newest pulse, on closer inspection, is composed of many smaller turrets on several more time and size scales. Although these are beyond the scope of the broader definitions for this storm type, they suggest complex motions whose significance is still largely a mystery. The system formed late, about 5 pm, went severe with a massive burst of new growth around 6 pm, then died abruptly at 9 pm. Despite west-southwest winds aloft, the rapid growth caused a sharp propagation to the southeast.

▼ 6.6 (W Texas, 25/5/85, 8:13 PM CDT, ESE, 52deg)

THE SUPERCELL

The supercell thunderstorm is the most powerful and potentially destructive local convective storm type known. It is responsible for a large percentage of the strongest and longest-lived tornadoes that occur, and almost always produces large hail, brief wind squalls and downbursts, and continuous lightning. It also inspires awe and fear by its sheer size and significant control over the local atmosphere as it sucks in and wrings out tons of warm, moist air.

This storm type is most likely when strong capping delays the release of energy until late in the day, then breaks down to allow one or a few isolated storms to flourish. The very efficient, jet-like updraft can exceed 50m/sec (55yd/sec), creating a continuous, overshooting dome above the anvil. The system is maintained by a dominant inflow which feeds air through the complex as a stream of short-interval updraft pulses. By comparison, the system as a whole evolves relatively slowly, remaining fairly uniform in appearance over a period of hours.

From a distance, the supercell's steady-state character can lull the observer into the false assumption that the cloud's interior is also changing little. From this perspective the backsheared anvil stands out against the blue, slicing back into the wind like a sharp wedge for 20 to 30km (12.5 to 18.6 miles). Downwind, the anvil sprawls as a thick, uniform sheet which may reach several hundred kilometres in length. A closer look at the storm's updraft

▲ 6.7 (Alberta, 27/7/80, 7:00 PM MDT, NE 94deg)

6.7 This storm is viewed from a 5.5km (3.4 miles) altitude with a wide-angle lens. It is a multicell type with a rather short pulse interval which has produced a long line of closely packed towers downwind of the core. The degree of separation apparent in this row of towers is quite surprising since the radar echo structure showed only a single, unified image. It points out the remarkable ability of storms to manage many different components and operate as a unit. The system's rearside shows several inflow streets converging at the updraft location where they rise and merge into it. The more recent and more vigorous growth, with crisp detail, is to the right of older and more diffuse glaciating towers. Above and ahead of the core are two levels of accessory cloud material. The higher one is a persistent pileus cap lifted from left to right by strong west-northwest winds at the storm's top. This also accounts for the absence of an anvil, which is only present on the far side. Below the pileus, there are also some thin shadowy streaks resembling a wave feature. The system formed to the lee of the Rockies around 4 pm, tracked southeast across the province, then died around 9:30 pm. It deposited a devastating hailswath for over 100km (62 miles), with hailstones exceeding golfball size and radar reflectivities over 60dbz for over two hours.

base reveals a short line of close, steeply rising towers (the flanking line) and where they join the main cloud, circular or laminar striations are sometimes seen below the base, hinting at the embedded mesocyclone rotation.

▲ 6.8 (S Dakota, 4/6/80, 3:30 PM CDT, S, 94deg)

6.8 This supercell storm shows a typically short updraft base (flanking line) in which clouds ascend to great heights in only minutes after first appearing. Since the line is also along the inflow/outflow interface, it is obvious that inflow is dominant. On the upper left, the towers are approaching maturity and beginning to spread out. Their outlines, faintly visible against the anvil, will blend into it and push back to the right, extending and maintaining the backsheared portion. The anvil is very high (around 15km, 9.3 miles) and shows shallow billows plus concentric rings at right corresponding to the separate outward

steps of its formation. The storm had produced three tornadoes over the previous hour and its top had reached 18km (11.2 miles). It began about two hours before from a cluster of smaller cells growing along a weak cold front (and coincident dryline) pushing east into much more humid air. Once this storm had become established, it became the only one in the region and all the other cells and clouds vanished. The system was shifting southeast, dropping 2–6cm (0.8–2.4in) hail along its path before weakening in eastern Nebraska during the evening.

6.9 ▶

6.9 Visual schematic of a tornadic thunderstorm, showing the wall cloud and its relationship to other cloud features (*Doswell, 1985*)

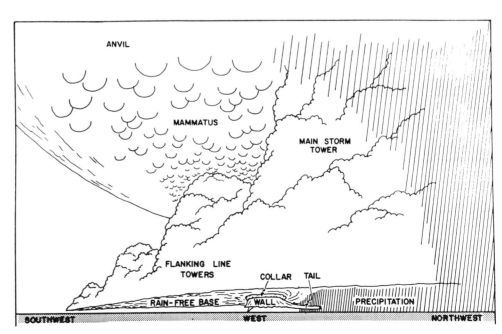

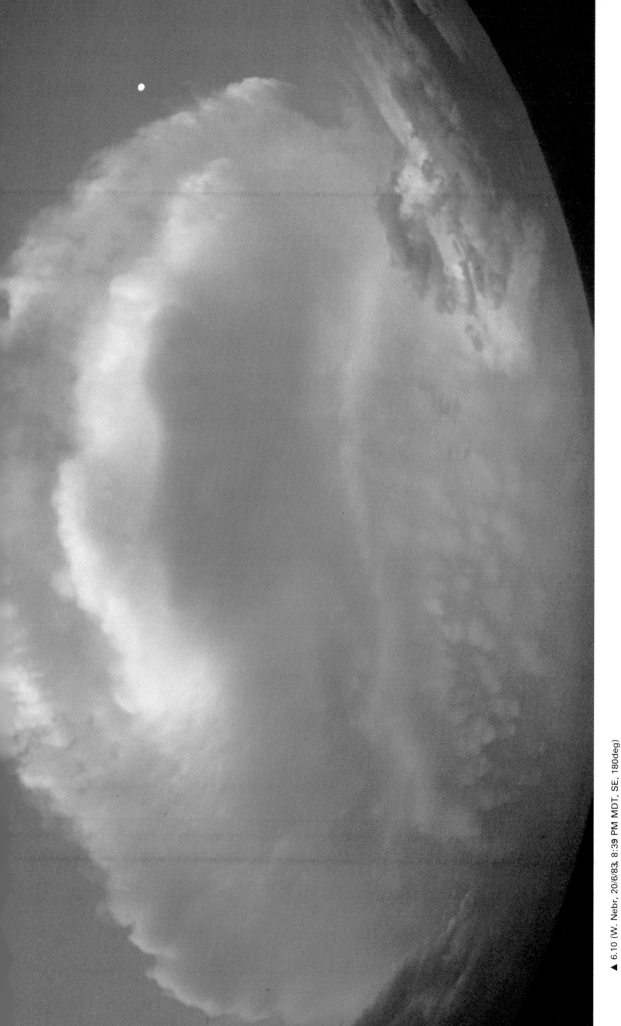

▲ 6.10 (W. Nebr, 20/6/83 8:39 PM MDT, SE, 180deg)
To get the 'big picture' you sometimes need a bigger view than usual. A near-fisheye lens was able to just squeeze this supercell rear view into the frame and although distorted, it gives a good indication of the immensity of these storms (notice the dwarfed full moon). The cloud extends from the zenith to the horizon and shows a variety of features common to this type: mamma on the northwest rearside anvil, a short flanking line lower right, layered rings of thick anvil from the spreading

updraft pulses, and a lowered area to the south of the precipitation. The system had weakened during the past hour but before that, had produced two brief tornadoes and a few 8cm (3in) hailstones. It began in eastern Wyoming around 5 pm where it sat in one place for two hours before drifting east. Residents in that area claimed they had nearly continuous hail and thunder for that entire period. The tornado event is shown and discussed in section Rotation and the Tornado, p 108

Among all the storm types, this one thrives especially well in strong vertical wind shear. Since shear would appear to be destructive to vertical development, this is a surprise, and suggests an important additional factor governing the system's survival. The rapid, multiple updraft pulses also couldn't be maintained locally without extra dynamic support. And indeed, the supercell contains a mesoscale rotating updraft (the 'mesocyclone') that streamlines and powers the flow through the system.

A typical supercell begins with a single, stronger updraft on the right flank of a regular multicell. A weak echo region forms in the updraft when precipitation is carried up and forward, free of it, and the first signs of the mesocyclone appear at mid-levels. The process is not fully clear but it is thought to form initially from the tilting owing to shear in the environment and be further enhanced by the spin-up ('ice-skater effect') as low-level winds converge at the column. When the updraft core collapses, the mesocyclone migrates to the area of strongest vertical shear between the updraft and a developing rear flank downdraft. The rear flank downdraft is colder and drier than the forward downdrafts, and originates aloft where the flow is deflected down as it encounters blocking by the updraft core (from 4 to 10km, from 2.5 to 6.2 miles). Assisted by evaporating rain below the backsheared anvil, the rear flank downdraft works its way down to the ground and seems to be related closely to tornadogenesis.

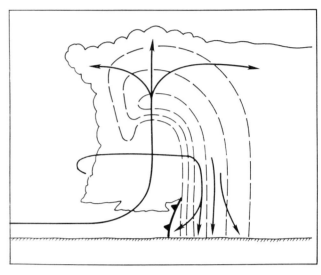

▲ 6.11a

6.11a Simplified cross-section of a supercell. The precipitation (dashed lines) forms aloft above a steep rainfree vault and is carried out ahead of the updraft region, where it descends in the forward flank. (*Doswell, 1985*)

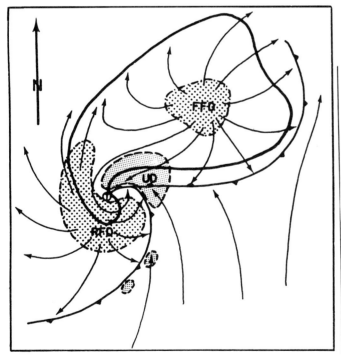

▲ 6.11b

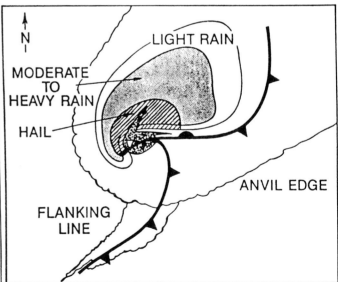

▲ 6.11c

6.11 (b and c) Top view of supercell. In (b), forward flank downdraft (FFD), rear flank downdraft (RFD), updraft (UD) areas and surface flow (arrows) are shown. (*Lemon & Doswell, 1979*)

(c) shows the precipitation and cloud boundaries. Cool outflow air is divided from warm air by a frontal structure that resembles the larger temperate zone cyclone pattern. (*Doswell, 1985*)

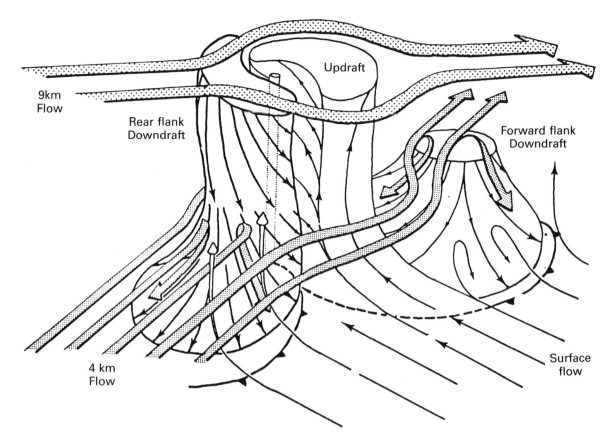

▲ 6.12

6.12 A three-dimensional schematic representation of a supercell in which the mesocyclone is fully formed. In an evolving supercell, this stage depicts a peak in intensity and temporary balance between the updraft core and an expanding rear flank downdraft. The mesocyclone (thin vertical cylinder) and tornado appear between these two airstreams. Flow lines throughout the figure are storm relative and conceptual only. (*Lemon & Doswell, 1985*)

The mesocyclone can evolve continuously, moving with the system, or regenerate in discrete steps as the main updraft weakens and reforms repeatedly. The entire system also can undergo a stepping process in which the outflow boundary surges ahead and around the updraft, pinching it off and forcing it to rebuild slightly forward. This is similar in structure to a frontal cyclone, with the gust front acting as a cold front that occludes the warm inflow repeatedly. When falling rain is drawn into the rotating column, it wraps around to form the familiar 'hook echo' on a radar image.

The south side of the supercell is reasonably well behaved and affords a good vantage point for watching the changes without much rain or low cloudiness. The flanking line usually has clearing sky behind it so it is easy to recognise where the storm's centre is. A modified version of the supercell has fallout behind the flanking line, as well as to the north and northeast out ahead of the storm. Between the two areas is a point, sometimes obscured by thick clouds and fallout, where the mesocyclone and possible tornado would be. The inflow curves into this point from the southeast along a long arcing path, causing the storm's structure to be easily and dangerously misinterpreted as either disorganised or weakening. This storm type is one that never fails to surprise and should never be taken for granted.

6.13d (7:36 PM, WSW, 34deg) ▶ pp88–9
(See p90 for 6.13a–c)

(Texas, 29/5/87, 7:03 PM CDT)

6.13a–g The storm environment is in a state of constant change. This series shows a few of the unusual features associated with an evolving cloud system on 29 May, 1987 in southwest Texas near the southeast corner of New Mexico. The storm, which began as a multicell, went briefly severe with large hail and a scrappy wall cloud. A large burst of outflow then forced the system into a weaker, steady-state period during which new cells developed gradually farther southwest to form an elongated, small line similar to a squall line in structure (see 6.15). The storm had formed on the dryline ahead of a disturbance approaching in the west–southwest flow aloft.

▲ 6.13a (7:03 PM, WSW, 110deg)

▲ 6.13b (7:11 PM, W, 80deg)

▼ 6.13c (7:36 PM, WSW, 34deg)

A thin line of showers in (a) divides an area of cool east–northeast outflow winds from the warm southerly low-level flow. Under the uniform anvil are low, darker clouds pushing to the southeast at right as a shelf but moving toward the system at left. The flanking line is still growing at its end but each new cell merely joins the elongated rainwall in this outflow-dominant phase.

In (b), the back end of the line has lowered somewhat, indicating that a better organised cell may be in the works.

The system has changed dramatically by (c). The old shelf had been slowly shrinking over the past half hour and is now a roll cloud that sits on the boundary between a weakening outflow and the prevailing south wind. The warm air is being lifted and curled down again as it meets the rain (note rain shafts curving around roll cloud). It attaches to a lowering which is the site of a relatively strong updraft. The fallout region is clearly distinguishable by its lack of detail or lower clouds.

The lowering, seen up close in (d) on pages 88–9 is quite a peculiar sight. It is a sort of variation on a typical wall cloud, with signs of a rotational component in the midst of a complex, twisted flow pattern and structure. Although the feature is moving towards us, all rising air at left is permitting it to 'grow' left across the flow.

By (e), the lowering has lifted somewhat as the updraft weakens. Air is still rising at left to produce strangely shaped clouds but the circular and discrete structure of the earlier cell is missing. The roll cloud is almost overhead and still moving slowly south.

In (f), what's left of the roll cloud now extends away from a circular bulge which marks the increasingly isolated, dwindling updraft core. Behind this 'tail', a lone inflow band focuses the rising air as it is being squeezed by outflow from both sides. Light rain and occasional thunder overhead accompany the dying storm, and the rain is set aglow by bright sunshine in clear skies a few miles away. For (g) see p 91.

We let the shower wind down and pass us. The sky cleared, leaving an orange anvil to the southeast. A wide view in (h) captures the brilliant finale to an amazing event. The cell which gave us those rare sights could be distinguished as a thicker, separate piece or extension of the anvil. Far below it in rain-cooled air, patches and strips of low stratocumulus were gathering to blanket the earth until morning.

▲ 6.13e (7:55 PM, SW, 24deg)　　　　　　　▲ 6.13f (7:55 PM, SW, 24deg)

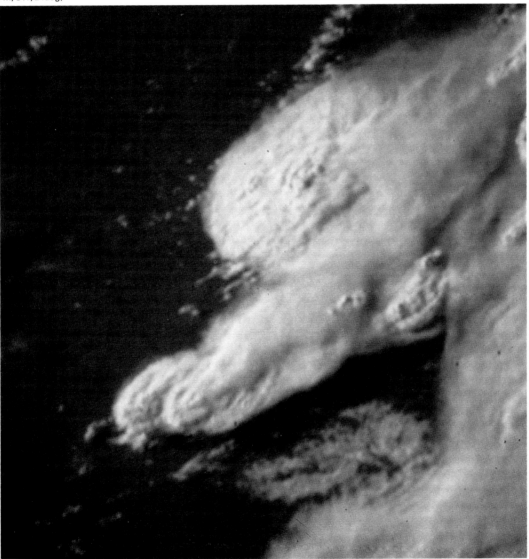

▲ 6.13g (Satellite picture 29/5/87 7:31 PM) This view is from just before the scene in 6.13a, at a time when the system was in its squall-line stage. The cloud complex has elongated along the axis of low-level convergence where the warmer surface flow meets a pool of cool outflow air behind (northwest of) the line. At right, the original cell has collapsed and a strip of smooth cloud shows where its uniform cirrus anvil is located. North of it, a new storm is developing rapidly and its grooved and sculpted appearance indicates active convection penetrating the anvil top.

At the lower left, the west end of the line also shows good detail. This is the site of vigorous convection also. The slightly circular pattern suggests a degree of organisation not yet visible from underneath (see 6.13a) but nevertheless indicative of the dramatic changes that followed. All along the southern edge of the line are low to middle clouds that flank the central rain area.

THE SQUALL LINE

As a severe storm event, the squall line is a stronger but generally smaller version of a typical, frontal squall line. Moderate rain may extend for many hundreds of kilometres along its axis but the active portion is usually less than 150km (93 miles) long. It often forms from the decay and transformation of a multicell, supercell, or cluster of weaker cells when an excessive outflow collapses the previous balance and begins to surge forward along a broad boundary. Individual cells form along the new interface and migrate up the line with the prevailing flow but line characteristics predominate over cell characteristics. The active core is along the leading edge where a sheet of warm air is being continuously scooped up into the system, condensed, then left in its wake as a large backhanging anvil. The scooping process may be nearly continuous, causing the system to propagate in small, steady increments, or it may be in larger, discrete steps resulting in a series of quite distinct, rapid forward moves with slower motion between. The embedded cells may be at any stage of development and are usually most intense right at the forward edge or near the line's southern end. Their intensity rarely reaches severe levels but the system's greatest impact occurs when the outflow surges produce damaging straight-line winds for brief periods along the ground.

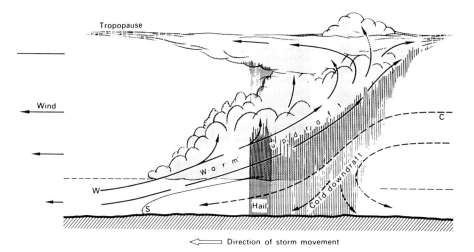

▲ 6.14   ⬅ Direction of storm movement

6.14 This is a cross-section of a self-propagating shower cloud in its simplest form. In reality, this simple two-dimensional structure is complicated and distorted by variations in wind speed and direction. The rising air, starting at W, is lifted by the advancing cold squall S, the cold air being below the continuous line. New cumulus growth on the left develops gradually into the main updraft which is inclined so that the rain from it falls out above the main cloud base into the air at C which is overtaking it. The cooled air forms the downdraft and inversion (below thin dashed line) at left, and spreads forward to regenerate the squall. If the squall makes a significant advance for a period, a shelf cloud may form below the cumulus base, as in 6.15; otherwise, cumuliform clouds will be more apparent.

▲ 6.15 (Texas, 29/5/87, 7:01 PM CDT, ENE, 110deg)

6.15 This photo approximates the situation depicted in 6.14. The storm is moving to the right and is preceded by a sloping anvil that leads down in steps to the wedge-shaped shelf. This low cloud is along the very forward edge of outflow, or above S in 6.14. The photo is of a system that became outflow-dominant for a period but could also represent other situations where the system survives by forward propagation, as in 'front-loaders' and most larger squall lines. Although cold air is advancing at the surface, warm air continues to enter the system above the shelf cloud.

6.17 A bright bolt marks the location of strong updrafts along the leading edge of an approaching storm. The older cell is in the background and its recent collapse has sent a burst of cool air forward, pushing a line of new towers up ahead of it, seen here as the dark area. Paralleling this, a line of low scud is boiling up near the ground where the outflow is rising abruptly into the developing storm. The bolt's dark surroundings indicate little rain falling there while the light at top is coming from the higher anvil.

▼ 6.17 (Texas, 16/5/86, 10 PM CDT, N, 47deg)

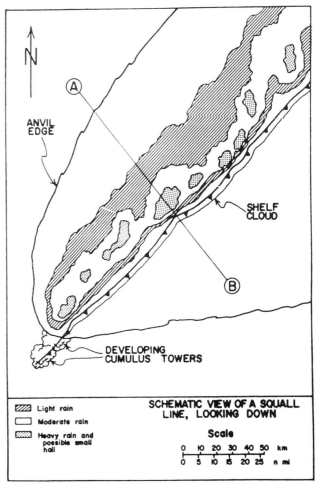

▲ 6.16

6.16 A schematic top view structure of a squall line showing a collection of cells crowded along the advancing squall front.

A typical cross-section along line A–B would appear as in 6.14.

The squall line resembles a cold front as its wedge-like forward edge cuts into the warmer air and migrates across the winds aloft. The leading edge gust front is marked by a long, low shelf cloud or arcus, plus fractus just ahead of the rainwall. At times, this fractus may lower significantly, forming a very ominous, turbulent mass that changes rapidly from moment to moment. These low clouds appear threatening and can be mistaken for funnels or other lowerings normally seen below the main updraft base of severe storms. During or just after the sudden wind shift, a strong squall of heavy rain and small hail descends on the area. But once this has passed, precipitation becomes steady and lighter under the widespread weaker downdrafts at the system's rear. Here, few low clouds are seen, the sky brightens slowly, rain tapers off, and the remaining anvil clears gradually. The forward edge continues, often surviving for many hours until the outflow either slows, weakening the inflow, or races out far ahead of the updraft zone, cutting the inflow off entirely.

Many storms either degenerate to a squall line or go through weaker phases that resemble the same thing, but over a much smaller area. If the system has a more or less permanent inflow and new growth region on its forward flank but does not extend for large distances as a line, it is a forwardly propagating type, or 'front-loader', which is discussed in Outflow Features, p 112. All these variations on squall lines have in common the arrival of new, developing cells ahead of a weakening rainy area. This means that hail and lightning bolts often precede heavy rain; features which can, especially at night, be the best available clues for assessing the situation.

OTHER TYPES OF SEVERE EVENTS

The three strong convective structures discussed thus far in the chapter are the most common but many other features, events, and storm-type variations exist which portray intense convective processes or severe effects. Some otherwise average thunderstorms can produce a brief, intense updraft with resulting wind squall, hail, or even a funnel. They are called 'pulse' storms and have been known to elicit false alarms and premature optimism from anxious skywatchers before they weaken again rapidly. When a cell within a weaker squall line collapses abruptly, a sharp outflow can bulge forward to trigger strong new convection along a curved boundary, known as a 'bowed squall line'. In other situations, strong updrafts can cause several kinds of funnels but these will be dealt with under rotation in Inflow Features, p 100.

Another variation on storm severity arises when a system remains in one location for an extended period, resulting in a local precipitation extreme and flooding conditions. Such events are uncommon because they require a near perfect balance between forward motion of individual cells and their maturation all at the exact same place. With all the changing variables at work, this is quite an accomplishment! It can most easily occur when a continuous airstream is lifted over high ground, forcing the convection to remain positioned there, or when cloud development is anchored to a stationary feature in the atmosphere like a front or zone of surface convergence. The system then also requires the updraft core to remain stationary, usually by having surface winds in a direction nearly opposite to those above carrying cells forward.

The dry conditions of the western high plains in the U.S. create storm effects not commonly seen elsewhere. High-based clouds with apparently unimpressive structures can, by extreme evaporation of their fallout, produce intense downbursts. Such events cannot be seen easily before they happen and are of considerable risk to aviation. Although a lack of moisture generally reduces the extent and strength of convection, there are several benefits to dry conditions for the sky enthusiast. Visibilities are excellent, ranging up to 200km (124 miles) and fewer low clouds add to viewing ease. The higher average bases of most cloud types give the sky a different perspective that opens it up, seemingly to extend forever as it curves down and beyond the horizon. Unhidden by haze or mist, sky emerges as an omnipotent presence reigning supreme over the landscape.

The Dryline Storm

Throughout the spring and summer in the Midwest U.S. the boundary between moist Gulf air and dry air originating west of the Continental Divide shifts back and forth with each passing system. The earliest convection on a given day frequently begins on this dryline then moves or propagates eastward later in the day. The boundary is often very narrow, with dew points jumping from below zero celsius to the 10–15 degree range in as little as 1km (0.62 mile)! The dry air is crucial to strong cells forming later in the day because it inhibits cloud development and upward mixing or dilution of surface heat and moisture until a critical point is reached. Once a storm forms, the denser dry air enhances buoyancy and evaporation to further drive the system.

The dryline storm is a spectacular sight, a mountain of creamy crunch surrounded by dark blue sky. It is structurally complete and may be a dry version of the supercell type. At the core is a narrow but powerful updraft below or around which laminar banding has often been observed. Rain falls well away from the updraft and hail is common, with large hailfalls or stone sizes being the most frequent severe event. A collapsing dryline storm can cause violent outflow winds, but otherwise outflow effects are minimal or non-existent. The system's updraft core distributes the air aloft and downwind in the anvil.

STORM ARRANGEMENT

During the day there is a tendency for convective clouds and storms to decrease in number while increasing in size and strength. The ultimate arrangement and types of storms to emerge are determined by the predominant factor at work in the atmosphere. Early (uncapped) development produces many smaller cells of about equal strength until, possibly, one or two rise above the rest. If the predominant factor is a surface-based linear motion, a squall line results. If an area is affected, a cluster will form and may mature into a mesoscale complex. The structures present also evolve over their lifetimes from one type to the next as their processes modify the near environment, setting up a series of reciprocal actions. When these actions act to enhance that environment favourably, as when a cluster of cells creates a local area of surface convergence, a much stronger system will emerge. Later, when this system's outflow begins to overtake inflow, it may decay to a weaker squall line again. Some multicells intensify to brief supercell-like status, and dryline storms can grow into other types when they move east into moister air. The system's actions can also modify its surroundings so as to prevent further convection, resulting in an early death.

System motion is usually in steps and closely tied to the shifting balance between inflow and the outflow surges. If the system moved entirely with the flow (not usual) air could both enter and exit at the front. In most cases there is a net imbalance forcing it to shift slightly to the side and this, combined with a wide range of propagation steps, leads to an almost infinite number of possibilities. In addition, the proximity of fronts or prominent local characteristics can also contribute to changes in motion by forcing new growth to occur selectively.

The best example to watch and study is an isolated severe storm, all by itself in a vast expanse of clear sky. This can happen when a strong inversion is weakened or lifted locally, causing the 'cap to break'. Once such isolated events are underway, they do not readily collapse and may continue until late in the night as a self-contained circulation that controls the environment for a hundred kilometres around.

The inflow and new growth pattern determines a system's motion. In the usual case this means a shifting to the right of the prevailing westerlies (this is reversed in the Southern Hemisphere) and these are labelled 'right-movers'. Another type, the 'left-mover' can result when new growth is on the downwind side of the surface flow. For example, a system under northwest winds aloft and south winds at the surface can have the updraft core on its northwest side, causing a shifting to the north. To overcome the disadvantages of such motion away from the inflow source, these storms usually travel at greater speeds and are as hard to track as they are deceiving to interpret. They can also produce tornadoes and other severe effects that come and go equally rapidly. Left-movers are more common as part of a splitting storm, where an original system's inflow slowly separates into two parts. Each one then carries on independently, with the right one becoming or remaining stronger. The left one usually remains weak or dies soon afterwards.

Storms can organise into clusters when similar conditions exist over an area. The cluster is easier to recognise from a satellite or radar view and has a collection of fairly similar storms separated by clear sky or thin anvil material. From a distance, several of its members could be mistaken for the separate towers of a multicell with a long pulse rate.

Another example of a larger-scale structure is the mesoscale convective complex (MCC). It may begin as a single storm or cluster but once underway, acts as a symbiotic system with separate elements interacting with each other in a more or less steady-state. In its early stage, the still distinct members can be severe. Later, it matures to a massive, uniform complex with an extensive anvil and embedded cells. During this stage, a large-scale weak rotation appears at mid-levels to maintain slow, steady convection. It eventually decays (sometimes a day later and a thousand kilometres away) into an area of light rain.

▲ 6.18 (W Nebraska, 30/6/83, 7:43 PM MDT, NE, 13deg)   ▼ 6.19 (Colorado 29/6/83, 3:29 PM MDT, S, 94deg)

6.18 This storm has a classic, sharp backsheared anvil with two banks of towering clouds on distinct inflow lines. About half an hour earlier there was only one, larger inflow area but this separated gradually as the original single storm underwent the splitting process. Notice that the left side is clearly less well developed and the right shows occasional overshooting tops, indicating that the stronger half resides on the right side. The system formed in late afternoon, moved northeast, then weakened by 10 pm after producing 4cm (1.6in) hail and funnel clouds at its peak just before the split.

6.19 A very heavy raincurtain identifies the mature cell in a storm system. This area is also brighter since more light is reflected from the dense sheet of raindrops. Dark clouds at left plus lighter showers, right, indicate that the system is propagating forward like the 'front-loader' type.

▲ 6.20a   ▼ 6.20b

(New Mexico, 24/8/83, 2:56 PM MDT, NW, 30deg)

6.20a,b This mountain-induced storm has become a small, self-contained system of continual growth and decay. Cells forming along the flanking line mature, dropping heavy rain at right which sends cool, outflow air down and back under the line. The outflow wind is accompanied by blowing dust which has reduced the visibility in a shallow layer along the ground. As this air moves under warmer air and up the slope at left, sudden lifting is forming irregular blobs of scud that rise to meet the next main cell in the series. The balance remains as long as the maturation rate exactly compensates forward motion.

In 6.20b only fifteen seconds after 6.20a, the scud has already changed quite a bit. The swirling, black mass is surrounded by various bits and threads of cloud material which are condensing in mid-air. Such strands of condensation are the result of upward acceleration of air which contains the stretched remnants of moist outflow which had been subjected to turbulent motions.

▲ 6.21 (Alberta, 27/6/81, 7:28 PM MDT, W, 12deg)

6.21 The storm environment up close is a fascinating, puzzling mix of shapes, light, and, perspectives. In this view, a distant bank of showers looks strangely out of place 'below' an altocumulus sheet. Actually, the sheet is closer and ends well before the cumulus which have grown in the warmed, clear air beyond the cloud deck. To the right of heavy rain, lighter showers remain while to the left, a small lowering is apparent; both suggest a propagation to the south.

6.22 The storm environment, seen from afar in good visibility, looks orderly and peaceful. The perfectly shaped anvil of a 15km (9.3 miles) high tornadic supercell distorts our perception of distance; it is over 150km (93 miles) away and about twice as far as the other cumuliform clouds growing 'under' it. They, as well as the altocumulus strips, are forming along the cooler ouflow which is being carried north by south surface winds. The scene is a sharp contrast to the complex, turbulent events going on under this cloud.

▼ 6.22 (New Mexico, 26/5/87, 3:39 PM MDT, SSE, 62deg)

▲ 6.23 (Texas, 29/5/86, 8:43 PM CDT, WNW, 46deg)

6.23 The storm environment is full of surprises. A few minutes ago, the sky was a neutral grey, with one brighter area separating clouds distant northwest from a large anvil on the north side of a rainy MCC to the south. Moist outflow, carried on southeast winds, was felt but not seen until it was slightly lifted by passing billows. The rolls were short and thin as they first formed above, then widened, thickened, and began to sprout as they moved northwest. They were almost as low as the few bits of fractus around, and evaporated in the distance where the wave train may have reached drier air.

(Kansas, 18/5/87, 7:33–7:45 PM CDT)

6.24a,b (a) There are two storm systems affecting this scene. The one distant east has towering new growth on its west side along the flanking line and appeared to be a left-mover from later evidence. This line has become extended by outflow from

a second, weakening storm to the south and is now moving north as a detached gust front, seen here approaching at right. Its forward surface has a laminar edge typical of a shelf cloud; yet it is quite atypical otherwise, being completely detached and moving in a different direction from usual. When the line passed, winds became strong southwest for about ten minutes and the air became decidedly cooler and moister.

In (b), the line has passed and continues north into completely clear sky. A remarkably straight baseline supports numerous cumulus towers that lean back into lighter winds aloft. Subsidence is breaking clouds up overhead. The line slowed, then stopped moving north about a half hour later and became a new line of thundershowers afterwards. These did not last too long, and were overshadowed by a rapid increase in activity to the southeast after dark which pushed east as a squall line ahead of a weak cold front.

▼ 6.24a (7:33 PM, E, 80deg)

▼ 6.24b (7:45 PM, WNW, 80 deg)

STORM-SCALE AIR MOTIONS
AND FEATURES

OVERVIEW

The best way to make sense out of the myriad of cloud features possible in the storm environment is to imagine the whole thing as one large interconnected three-dimensional airflow pattern. By putting these 'mental streamlines' on the view, the separate parts join up in the process and our attention is more easily focused on those active places where rapid changes are most likely. This approach is just as valuable for other sky events, including the multitude of cumulus towers that precede a storm's development. Once we are tuned in to the scene, armed with a little basic information and experience, the key elements or changes stand out naturally from the busy background.

Even before serious convection begins, cloud layers should be studied for clues to that day's conditions and the likely results later on. With flat, uniform cumulus under capping conditions, serious convection will be late and will first appear in a favourable location. But once seen, this storm will be the one to watch. Early convection is messier and the first cells may give no true indication of stronger activity later on. In fact, they may cool the atmosphere nearby, making that area the least likely place instead. Convection and the broad horizontal flow will affect each other so there is usually a link between what's happening now and what will occur later. New growth along old outflow boundaries, the lining up of developing towers, and the drawing together or clearing out of clouds in one area all indicate the emergence of a dominant storm system.

Although convective updrafts are confined to a relatively small area, they distribute the air vertically, then horizontally. The horizontal aspect is not well understood but its widespread nature causes many secondary effects, some quite subtle. One common example is the way a single storm can clear a large area around itself (especially at the back) of all other clouds, including cirrus. The central updraft induces a broad subsidence to accomplish this and the sudden disappearance of clouds usually means nearby convection is becoming stronger and better organised. (For an example of subsidence producing 'ghost anvils', see the series in 5.3.)

The simple view of a single cell storm portrays a central updraft surrounded by downdrafts. This is then complicated by wind shear which puts a tilt on the system. Outflow moves downwind of the inflow and descends with rain-cooled air. The 'simple' picture is becoming less so, since this air can either inhibit new convection (negative buoyancy below cloud base) or promote it by cooling the layer above the main cloud base. In most storms the former is more likely but in situations where warm surface winds are entering from the east or north, the latter can be important. Regeneration and propagation are a big part of survival so the delicate balance along the inflow-outflow interface is another complication. A storm fully under way is not affected much by mixing or entrainment of air external to the system but can be dramatically altered

if the inflow is smothered or distorted by the advancing outflow. When several storms are involved, each with their own special contributions to the environment, the picture is almost impossible to analyse accurately.

With storm systems the connectedness to the surrounding conditions and features, even over large distances, is fundamental to our understanding and appreciation of them. Peculiar cloud features or microscale details belong to the process as much as the more obvious signs of 'controlled' air motion. Even apparently unrelated parts of the wider scene may be linked by motions and changes not easily recognised. The collapse of distant showers or the oscillations of a storm's top through the tropopause can set up wave motions that may travel great distances before becoming visible in other cloud patches. Similarly, the presence of high land upwind in the low and mid-level flow can twist or alter it far from the source, causing convection to begin or proceed selectively. There is little known about either the host of special cloud forms or the relationships between them. Exact causes are still unclear and subject to much speculation. The attentive observer can discover many fascinating things in the storm environment and may even acquire an edge on current knowledge.

INFLOW FEATURES

The inflow to a storm complex is usually in the south quadrant except with squall lines or other forward-generating systems, when it would be on the east side. In this area, the cloud base is fairly uniform, dark, and rippled or textured but with little rain falling from it. It extends up to the main, mature cell where the heavy fallout begins. From a distance, this updraft base (sometimes called the rainfree base) is very flat and extends away from the storm as the flanking line, a line of cumulus towers whose growth is being aided by surface convergence ahead of the outflow region. As newer cells mature, streaks of rain can be seen forming under the base to extend the older raincurtain in the direction of propagation.

Air approaching a storm ascends gradually from many kilometres around, then more steeply after passing through the condensation level. A certain amount of mixing occurs of the low-level moist flow as it rises through drier air, and this will affect the height and uniformity of the base. If moisture increases the base will lower, and there are three common ways this can happen: the system updraft can intensify, causing the air to rise more smoothly into the cloud with less mixing en route, the rising air can be moistened by evaporating raindrops or entrainment of moist outflow air, and the air entering the updraft may become intrinsically more moist either from passing over wet ground or by drawing on an increasingly moist flow into the region. Any or all of these modifications will be reflected in the base detail, providing the observer with an accurate barometer on subtle changes in the system's circulation and strength.

When the system becomes severe and begins to control the airflow nearby several inflow bands may form.

▲ 6.25 (Texas, 27/5/87, 7:22 PM CDT, NE, 80deg)

These are long, thin, horizontal lines of irregular or laminar cloud elements that lead into the core from the southeast along a gently curved path, sometimes dozens of kilometres long. Like cumulus streets, they indicate local convergence in the flow.

Another change occurs under the updraft base about 2–5km (1.25–3 miles) from the rain area when a lowering extends down from the main cloud base. It may be smooth or have bits of forming fractus hanging down below it. The lowering can assume a stepped appearance, especially from afar, with a flared, wedge shape. This is called a pedestal cloud and denotes an intermediate step in its development towards a full-fledged wall cloud. The wall cloud is more nearly circular and can lower to within half a kilometre of the ground. It forms at the base of the mesocyclone and rotates with it. A fully formed wall cloud is not a common sight and may only appear briefly before losing its compact structure. In the supercell it is more persistent, and more likely to have funnels or tornadoes emerge from it. In rarer situations, there may be laminar bands around the wall cloud or a ring above it, known as a collar cloud.

6.25 This is what a typical multicell storm looks like from the inflow side. The darker flanking line is also acting as an ill-defined shelf cloud (note ragged lowering under it) since the outflow is temporarily pushing to the east. Between the line and the rain area distant left is another lowering, which is where a wall cloud would form if the system intensified. This did happen about an hour later, but only after repeated attempts were checked by persistent outflow surges. Surface winds were gusty south and cells moved slowly northeast. This storm began as the one pictured in 5.20.

These various accessory clouds are all manifestations of a powerful updraft which is drawing air in from all sides over a small area. They consist of cool, moist (often outflow) air which has been forced to rise locally. Their short life, transient nature, and small size suggest that optimal conditions for their presence are quite rare and evanescent. This part of a storm harbours many other unusual cloud features and affords us a front-row seat at the battleground of a cold and warm air struggle.

▲ 6.26 (Texas, 15/6/85, 9:10 PM CDT, NW, 30deg)

▼ 6.27 (Texas, 28/5/87, 5:55 PM CDT, SSE, 94deg)

6.26 A finely textured shelf cloud base precedes a wall of rain set afire by the sunset light from the clear skies behind this weakening storm. Within the line, pockets of regeneration show up as dark, straight rainshafts. In other places, rains have thinned and the streaks exhibit a great range of angles and textures in response to fine, local variations in lighting, droplet size, and air motion. They are inducing numerous separate downdrafts that become distorted as they impinge on the ground.

6.27 A sprawling flanking line stretches across the view. It is attached to a storm moving away to the northeast. The tops of the towers are shearing forward in south winds aloft but the line is shifting slowly east. A closer inspection of the base shows that it is not perfectly straight and continuous, but more like a string or group of individual cumulus elements. As confirmation of weakness, the line abruptly evaporated half an hour later, leaving behind only a peaceful cirrostratus anvil.

6.28 (Texas, 16/5/85, 12:35 PM CDT, SW, 74deg) ▶

6.28 The sky is in a state of rapid change. A heavy storm is passing by to the northeast and has spread a blanket of cool air to its south. At the boundary of the cool air and warm southeast winds, a long convergence line has formed a cloud street along which air is being funnelled into a new area of updrafts. The smooth top on the cloud street suggests that the lifting is due to air forced against a stable layer which could have formed above the outflow. General sinking motion is also evident from the crinkly mamma. A low strip of scud indicates a second axis of lifted air. Its recent formation is a sign that the developing new cell, though still innocent in appearance, is already beginning to shape the local circulation. It became a severe storm with golfball-size hail before an hour had passed. Cells were moving east, with propagation to the south.

6.29 The use of a fast filmspeed has permitted these cloud details to be captured using soft light from several distant lightning flashes. The lowered area is a spidery looking wall cloud which has, on close inspection, a funnel pointing down from it. The point is blurred from slight movement of the cloud between flashes. Lenticular patches at left are forming in stable air above shallow outflow, possibly under the rear flank downdraft. The main inflow and flanking line entered the system along a steady surface flow from the southeast. The storm is a supercell which travelled a 500km (311 miles) path to the east-southeast (cells were moving east-northeast) between about 5 pm and 4 am, from its multicell beginning in northeast New Mexico to its squall-line demise in central Texas.

6.29 (Texas, 25/5/85, 11:30 PM CDT, NNE, 54deg) ▼

▲ 6.30a (5:58 PM, WNW, 46deg)

▲ 6.30b (6:13 PM, WNW, 46deg)

6.30a–d (Texas, 25/5/87, 5:58–6:35 PM CDT)
Several steps in the changing cloud features under a multicell, dryline-type storm. It formed along a northeast-southwest line of storms just east of the dryline, with mostly clear skies northwest and southeast of the line. Surface winds were gusty south, and cells moved rapidly north-northeast. In (a) the flanking line shows developing towers leading up to a wall of rain. The curved rain edge shows that the system has a sloped outflow/inflow boundary under the line. After moving north with the system, we now see (b) that the flanking line has shortened but become more compact. Its base has lowered a bit, looking slightly like a pedestal, and its right edge (faintly visible through rain) is being shaped by the beginnings of rotation.

▲ 6.30c (6:27 PM, WMW, 80deg)   ▼ 6.30d (6:28 PM, WNW, 110deg)

We have driven north another short distance, and now (c) the updraft core has taken on the shape of a laminar cylinder as rotation works down closer to the ground. It is still only weak rotation and probably does not signify a true mesocyclone. It is also not a typical wall cloud but, rather, a high-based dryline variation of one. The updraft is surrounded by rain and downdrafts which have helped to set it apart. The column, looking like a stack of tyres, is drawing stable air in at the sides to cause the laminar edge.

The updraft column is between the two rain areas. The strip of scud bits joins the precipitation core to the updraft by a thin channel of cool, moist air moving left. Some scud also formed under the column at times.

By (e), the lowered area has passed rapidly by and lost much of its compact structure in the last few minutes. Bases are lifting, rotation seems to have stopped, and the outflow is slowed at the weakening updraft but still passes under it and beyond. Surface winds have become northeasterly at this point, verifying that the outflow-dominated decay stage of the system is now in control.

▲ 6.31 (Colorado, 19/5/85, WSW, 94deg)

6.31 The menacing, wrinkled base of an old-fashioned high-plains hailstorm fills the scene. In the centre is a lowered pedestal cloud situated under the main updraft. It slopes up and to the right, and its upper region is brightest where a vertical surface reflects light better. Around the north and west sides of the core, heavy rain and hail are falling, and they are producing a northerly outflow that is slanting the rainshafts forward. This air is then circling around to the storm's south and helping to maintain the focused inflow from the southeast. Bases are relatively high and lower clouds are absent in the dry, unstable air mass. On a pass under the pedestal around 5 pm, hail larger than 3cm (1in) was encountered. Cells were moving east-southeast but the system was propagating to the south-southeast.

6.32a,b (Texas, 22/5/87, 6:36–6:41 PM CDT, W, 80/62deg) A narrow but energetic multicell thundershower is sending one tower after another up from a short flanking line. Steady northeast surface winds just north of a stationary front become light, then west-southwest above cloud base, resulting in an elongated storm whose core may remain almost stationary for long periods. The thin, distant strip is below and probably behind the system, and is the remnant of an earlier roll cloud that moved southwest away from the rain area. The angle of the updraft base (from south-southeast) shows that air entering the system has been twisted around by the opposing flows below and above the line. See also 6.32b.

The closeup in (b) a few minutes later shows a developing pedestal cloud. Air is rising here, then curving to the right where fallout is evident. The lowered, slightly laminar 'tips' of the pedestal are typical, but no clear explanation exists for its concave centre. The cloud formed and disappeared several times during the life of the storm but never became very pronounced.

▲ 6.32a

▼ 6.32b

▲ 6.33 (Texas, 25/5/87, 8:11 PM CDT, NW, 62deg)

6.33 The updraft base of a severe multicell stretches across the view from heavy rain at night. The lowered part is a ragged wall cloud trying to take shape. It showed weak signs of rotation and varied from second to second as its base descended then rose again in response to slight changes in the mesocyclone strength. After a short while the wall cloud weakened, possibly due to competition from another, newer storm just west of this one (rain lower left, see 6.34).

6.34 Another ragged wall cloud looms a kilometre or two ahead. It formed in the last of a long line of storms and is now being fed by an undisturbed inflow from clear skies all around the south side. Some signs of cyclonic rotation can be seen from the scud which is rising and joining the wall cloud like curved flakes around its rim. The low area had migrated slowly back southwest along the flanking line, formed a brief funnel, and is now gathering itself together and shifting more quickly northeast again. A few minutes after this photo, it became a large, circular, fully formed wall cloud to the northeast that coincided with the onset of a supercell phase in the system. A tornado appeared under this one later, along with many others in the region that evening.

▼ 6.34 (Texas, 25/5/87, 8:41 PM CDT, N, 80deg)

▲ 6.35 (Wyoming, 13/6/86, 5:43 PM MDT, N, 18deg)

6.35 Part of the flanking line (a closeup view) of a storm to the right has begun to lower as the axis of maximum updraft is shifted left by increasing outflow. If the outflow continues to push left, the lowered area will shift too, but if the updraft can gain enough control the present lowering could develop quickly into a wall cloud. In the meantime, the scud formed only a minute ago and is rising into the base. As it turned out, outflow prevailed and the base remained ragged and disorganised. The storm was moving east and had produced several brief funnel and tornado sightings over the previous hour but only weak rotation thereafter. The region was in a very moist upslope southeast flow.

6.36 The photo shows a closeup of scud and stringy cloud fragments forming in mid-air below the updraft of a storm to the right. Outflow rises along the sloped point on the left edge and joins the inflow there. The lowest parts nearly reached the surface in very moist air above water and in the wake of the rain. The system was moving east-southeast and was part of an area of convection associated with a weak trough passing through a cold-low situation aloft. On the same day over Lake Erie numerous cold air funnels and waterspouts were reported.

▼ 6.36 (Ontario, 28/6/87, 4:30 PM EDT, WNW, 18deg)

ROTATION AND THE TORNADO

The atmosphere is filled with examples of vortex motion of every conceivable size, from continent-wide cyclones down to the tiny swirls of dust on a hot roadway. All of them share with us a simultaneous mix of amazing simplicity and astounding complexity.

Ordinary updrafts contain no rotation. As they become stronger and have an organised inflow leading up to them, slight rotation may exist. Once initiated, this rotation may accelerate by the conservation of angular momentum as air approaches the updraft column. The centrifugal force lowers the pressure in the centre, causing moist air to condense below its usual level in the form of a small funnel. This 'simple' type of rotation accounts for most funnel occurrences, including those associated with waterspouts, cold-air funnels, and weaker convection. The rotation is almost always cyclonic.

Early thinking about tornadoes assumed that they formed from the tightening-up of simple rotation around a narrow vertical axis. More recent research suggests a different mechanism operating within a supercell storm. The mesocyclone begins as a rotating updraft. The rear flank downdraft develops near the updraft and the mesocyclone is transformed. Rather than a rotating updraft in which the centre of rotation is near the centre of the updraft, the mesocyclone's centre of rotation comes to lie on the interface between updraft and downdraft. Thus, updraft and downdraft end up rotating about one another. This process begins most often at middle levels within the storm, developing downwards as it proceeds. The mesocyclone descends (seen as a wall cloud) and the

tornado forms near its centre, touching down near the updraft-downdraft occlusion point on the ground. The exact method whereby a violently rotating vortex can emerge from the larger mesocyclone is not known, but what is known can be summarised in the following way:

> Once one has a storm and a mesocyclone within that storm, there are various candidate processes for producing the rotation of the tornado. There is not presently a consensus about which process (or combination of processes) is the correct one and the truth may be that any or all of these can produce a tornado in some situations. The candidates include: (1) local increases in low-level convergence into the updraft, (2) generation of enhanced rotation about a horizontal axis along the forward flank outflow boundary which is subsequently tilted into the vertical as it is pulled into the updraft core, (3) tilting of horizontal rotation produced within the inflow region to the updraft, or (4) complex dynamic processes associated with the mesocyclone which are not completely understood.

(Personal communication, Charles A. Doswell III, NSSL, Norman, OK)

It was once thought that all tornadoes had to rotate cyclonically but there have been a few that did not. Most tornadoes have wind speeds in the 150–300kmh (93–186mph) range and diameters of 100–500m (110–550yd); it is remarkable that air can be made to move at such speeds over a small area.

The tornado funnel is made visible by condensation due to extremely low pressure within it. Under varying circumstances, air may ascend throughout the funnel or just up the sides, with weaker descending air in the middle. In general, the most violent tornadoes have wider trunks

▼ 6.37a (7:12 PM, NNW, 52deg)

with a downdraft in the middle reaching the ground. There may also be several secondary vortices revolving around the main trunk. Once the mesocyclone begins to weaken or collapse, the funnel becomes narrower, elongated, or extended diagonally before it lifts and shrivels in its final rope stage. You can imagine that the balance between a pair of opposing airstreams bound tightly together is a tricky one to maintain for very long!

Another more common tornado type is the gustnado, which forms on a gust front due to stresses induced by the uneven forward motion of a strong outflow surge. It is small, relatively weak (F0–F1), short-lived, and not usually accompanied by a condensation funnel.

Weaker tornadoes associated with some severe multicell storms are sometimes called landspouts because they resemble waterspouts. With these, there is no significant mesocyclone but still enough rotation and stretching of vorticity to form the embedded vortex and extend it to the ground. Cold air funnels and waterspouts are common in cold, moist air masses that contain strong convection but weaker shear. They can be seen under rapidly growing cumulus or the individual cells of a weak cumulonimbus.

▲ 6.37b (7:14 PM, NNW, 30deg)

▼ 6.37c (7:20 PM, N, 30deg)

6.37a–c (W Nebraska, 20/6/83, 7:12–7:20 PM MDT) The three photos in this series depict several steps in the life cycle of a tornado (estimated F2) in far west Nebraska. The storm was a classic supercell (see also 6.11) which remained nearly stationary for several hours, then produced two tornadoes as it began to slide east. The first was a brief touchdown, but minutes later the second one, shown here, stayed down for 3–4 minutes, destroying much of a farm and uprooting several large trees. Surface winds were southeast and a pocket of instability (short-wave trough) was approaching in the west-southwest flow aloft. Although this storm was the only one for several hundred kilometres during the early evening, a rapid increase in new convection after 8 pm brought a restless night filled with many other tornado reports.

The light, disc-shaped cloud in (a) is a well-developed rotating wall cloud. In its centre a fat funnel reaches part way down; it is incomplete since the relatively dry western air is mixing into the circulation to deter condensation closer to the ground. It is already a tornado, however, as shown by the distinct debris cloud below it.

After a few minutes the usual smooth, tapered funnel is observed through a zoom lens in (b). This time debris is not visible, indicating that the tornado has lifted off the ground even though the funnel is now lower. The already clearing sky to the west is rendering it a ghostly white, and several tufts along its perimeter suggest smaller-scale air motions are embedded in the rising airstream. The main rain area is well ahead and to the right but lighter showers are lingering in the broad backhanging cloud deck as well.

This part of the storm is now weakening, (c) with the wall cloud deteriorating into ragged pieces. The funnel is shrinking into a thin, contorted, rope-like tube and retreating into the cloud base. During the whole time it was visible, there was no sound, little wind, and only the odd, distant rumble of thunder, giving the illusion that nothing of consequence had happened nearby.

109

▲ 6.38a (5:10 PM, WNW, 62deg)

6.38b (5:12 PM, WNW, 40deg) ▶

▼ 6.38c (5:11 PM, WNW, 18deg)

(Colorado, 8/6/86, 5:10–5:18 PM MDT)

6.38a–e A five-minute F2 tornado event on the northeast side of Denver is presented in the photo series. The tornado caused extensive damage along a 50m (55yd) wide, 2km (1.25 miles) path and formed just after the intense phase of a high-based storm which had moved east of the mountains. A southeast surface flow coupled with southwesterly winds aloft in advance of a strong upper trough had set the stage for a classic severe outbreak on this day.

We were quite amazed to see this distant debris cloud (a) at the ground without any visible funnel above it. A closer look at the updraft base (dark line, centre) reveals a small, ragged, but rotating wall cloud which shows up as a light, fuzzy area below the base. The parent storm is to the right, with the nearest rain about 5km (3 miles) away.

(b) is a closeup of the debris cloud. It is being drawn up the outside of the tornado vortex, permitting it to be seen despite the absence of a condensation funnel. The tornado is probably at its peak intensity here.

The entire vortex is becoming faintly visible in (c) by the rising debris, and the hollow appearance is good evidence that the air rises primarily up the outside of the column. The tornado is weakening at the ground because the debris cloud is now smaller. A bright sky filled with distant thunderheads provides an ideal, surprisingly peaceful backdrop for the delicately shaded vortex tube.

In (d), the vortex has now definitely lifted off the ground and is beginning to narrow at its bottom. Debris has largely dispersed, but what remains is still rising and has almost reached the cloud base at the top of the photo.

The last photo (e) is a very wide view of the entire

▲ 6.38d (5:14 PM, WNW, 40deg)

cloud system a few minutes after the tornado had vanished. The darkening sky at left is another storm on the same axis as the main one, distant right. The elongated remains of the wall cloud region show up as a lighter cloud face which ends abruptly on the right side. A closer look at this part of the sky reveals cloud patterns indicative of southeast low-level winds rising and curving to the right in the anvil.

▼ 6.38e (5:18 PM, NNW, 110deg)

▲ 6.39a (6:58 PM, NW, 110deg)

OUTFLOW FEATURES

The greater part of an average storm is occupied by outflow air. Most of it spreads out from the precipitation area towards the east and north, with some also spreading back to the west. If the outflow is confined to a shallow layer near the ground, there may be turbulent-looking clouds above it or a sky of broken cumulus that are all about the same size. If the outflow fills a deep layer, as might be expected in steady rain, the main cloud feature is a solid, uniform altostratus or nimbostratus sheet with ragged bits of lower cloud under it which may, at times, thicken into patches or layers of stratus or stratocumulus.

The outflow air is cool but often more moist than the inflow because it has evaporated raindrops during its descent. We may be tempted to disregard its effects, particularly in the rainy areas, but if we did we would miss an entirely separate group of cloud features that are rarely given any attention. The cool, moist layer is

very sensitive to the slightest changes in temperature or local lifting and such changes are immediately visible in the thickness, location, and transitory elements of these low clouds. A few moments of warming sunshine and, from nowhere, fractus begin to fill the sky. A little lift over nearby hills, and wads of stratus gather above them. These effects can be very localised, or blanket a large area following fluctuations in the circulation of the entire storm. The finer details of the rapidly forming and vanishing pieces of cloud are fascinating to watch, and reveal microscale processes that still confound our comprehension.

The other part of the system where outflow affects cloud appearance is along the boundary between the warm air and the advancing cooler air. If this boundary is stationary or diffuse, the updraft base extends up to and ends where heavy rain begins. If the outflow is pushing forward, a sharp, sloped interface exists at which cooler air is ploughing forward, forcing the warmer air to rise up the slope. This produces a shelf cloud (or, arcus), a long, wedge-shaped, low, cloud bank with a flat base and smooth or laminar upper (forward) surface. The smoothness is due to the forced lifting of stable air ahead of the advancing cloud bank. The shelf cloud base is lower than the updraft base and its top usually merges with it. It can span the length of a storm or be confined to a small area around the rainwall, but always indicates a steady forward motion of cooler air into warmer air. In a few cases, the shelf and updraft base may be merged or synonymous with each other if strong convection is under way at the interface. This happens in squall lines and 'front-loaders', a nickname for storms that have all their new growth on the forward edge. When a shelf cloud shrinks or vanishes, it means that outflow has slowed down or become diluted due to extensive mixing along the interface.

▼ 6.39b (7:09 PM, WNW, 62deg)

6.39a–d The following four photos refer to the same general system. After leaving (a), we drove east out ahead of the charging outflow to take (b) and (c). The last photo was taken after outflow again caught up and passed our position.

Light east winds are about to become strong, gusty west winds as an outflow surge from a heavy hailstorm pushes this low, churning shelf cloud forward. The pronounced lightening of the sky behind the shelf but just ahead of the rainwall could have been caused by sinking air in the wake of the sharp gust front. See also 6.39c and 6.39d.

Below the white anvil in (b), clouds are layered in successively thicker and lower steps to the shelf, which surrounds the strongest cell at lower right. There are two stages in the outflow, seen by the two lighter, lowered edges at left. The unusual pink-brown tint is reflected warm skylight from along the opposite early evening horizon. The area to the right has been under cool air for a while so its clouds have settled into more stable stratocumulus rolls.

(c) is a closer look at the left side of (b). The older, distant cell has weakened but a more recent new cell at left had formed southeast of it on the outflow boundary and has already added its own portion to the shelf. Cells moved east but the system was shifting sharply southeast by propagating forward every 20–30 minutes on the south end of the outflow boundary.

In (d), outflow pervades the scene. High above, a soft, light altostratus anvil spreads silently forward from a distant storm. Below it, a host of disorganised clouds at several levels are the temporary reminders of a passing, weakening flow of cool outflow air. The shallow layer of cool air is moving left and cloud elements that are forming in it rise, then hang back in the lighter winds above the layer.

▲ 6.39c (7:14 PM, SW, 80deg)

▲ 6.39d (7:35 PM, S, 62deg)

113

◀ 6.40 (Texas, 29/5/87, 3:25 PM CDT, ENE, 34deg)

6.40 A very large, high anvil with faint mamma is overspreading the sky from a big storm in mountains to the southwest. While a mild flow from the southeast continues at lower levels, fallout has cooled the layers below the anvil, making them more unstable. The cooling is aiding small cumulus to sprout into castellatus. A few distant scraps of lower cloud lean to the right in a light surface flow dictated by the local terrain.

◀ 6.41 (Iowa, 1/6/87, 7:51 PM CDT, NW, 80deg)

6.41 An irregular shelf cloud is forming clear ahead of a line of heavy cells. The cool air has not moved far ahead of the rain so the interface is steep, with thick scud rising to join and lower the shelf base. The shape of the lowered dark area suggests it is at the intersection of separate outflows. It developed into a strong new storm within half an hour. Cells were sliding northeast along a slow-moving cold front.

◀ 6.42 (Nebraska, 17/5/87, 7:27 PM CDT, SSW, 80deg)

6.42 A weakening thundershower at right has sent out a cool outflow which is pushing out and away as a curved shelf cloud. A high, finely textured base slopes down to contrast with the low, ruffle-edged shelf. In the distance, clear sky south of the system is faintly visible, but it is obscured at left by a few newer showers that formed where the outflow had lifted warm, humid air along the southeast flank. These showers are an example of (weak) forward propagation of the system. The outflow moved ahead for another half-hour, then weakened as it mixed with warmer air well away from the eastward-moving precipitation area.

It is interesting to consider that the sudden, rapid appearance of a shelf cloud means the storm may be collapsing or entering an outflow dominant phase. The same is true for the flanking line, which may lower and become a gust front for short periods when outflow air gains temporary control and accelerates forward. A little while later, the line may slow again, in which case bases will rise, and the flanking line will re-establish itself until the next outflow surge. If the storm retains its strength, this process may repeat several times. If the system collapses, the cool air may flow a great distance away from the original core and be preceded by a very long gust front which eventually separates from the dying storm. If it then slows, it will either dissipate or continue to develop new convection which may lead to a separate storm far removed from the previous one. The gust front is usually followed by clearing skies, because drier air from above the outflow is mixed downward in its wake. The same process can be seen behind the leading edge of an extensive shelf cloud, but in this case the downward motions within the thicker cloud layer fail to clear it. Instead, the sky takes on a chaotic, choppy appearance in which many dark elements are interspersed with small, bright gaps in the slowly subsiding air.

▲ 6.43 (Alberta, 7/7/82, 8:27 PM MDT, WSW 74deg)

6.43 The collapse of a storm to the left has sent cool air racing north in an already strong south flow, forming this shelf cloud. The front surface is multi-tiered as stable air is lifted and shaped by the sudden surge. If we imagine the onrushing air pressing forward, the effect is not unlike the wrinkles that form in a fluid ahead of a compression force. After the line moved by and slowed down, rapid new development began, bases lowered, and a line of storms soon occupied the northern horizon.

6.44 There are several puzzling aspects about this double shelf cloud. It is being produced by outflow, but the source of such cool air would appear to be right of its centre, where showers and an older cell are located (the rain above the tree is from a new cell just formed). Cool air is moving southeast but somehow the shelf has become curved, as if an invisible source were centered in the distant bright area. The doubling has no explanation yet, but the gap must be a level at which air cannot condense, suggesting either a thin stable layer or zone of slightly drier air between cloud base and surface outflow.

▼ 6.44 (Alberta, 15/7/80, 7:50 PM MDT, SW, 94deg)

▲ 6.45a (6:20 PM, NW, 94deg)

6.46 This is a shapely shelf cloud and gust front combination. The main cell is distant northeasterly, and in its wake a large reservoir of cold air is pouring south. The air is dry and most showers have already evaporated behind the line. Along the front edge several steps mark successively lower bases, with each one reflecting light differently depending on the angle of its surface. At the bottom is a rusty-brown rim of ragged clouds where air is rising into the base. The light hole could be due to some subsidence behind the line. Cell motion was southeasterly but the gust front continued south and regenerated into a long squall line that persisted all night.

▼ 6.46 (E New Mexico, 26/5/86, 7:08 PM MDT, ENE, 110deg)

▲ 6.45b (6:20 PM, SSW, 74deg)

(Alberta, 15/8/82, 6:20 PM MDT
6.45a,b A pair of photos captures adjacent parts of an approaching storm complex. The scene in (a) is the southern portion of the same system presented in (b). A two-tier shelf cloud is pushing out ahead of a collapsed older cell whose rains are thinning (distant, lower right). On the right, outflow is weaker and the shelf base remains flat and uniform. At left, the outflow is rushing into warmer air and has formed a lowered lip on the forward edge of the shelf. The dark rain area is a new cell that has just formed, indicating that the system has propagated forward in a large, discrete step here. However, the graphic rain shafts on its forward edge suggest that the big steps are separated by periods of continuous propagation ('front-loader') along the sloped inflow/outflow interface. The tail is probably due to some low-level convergence on the boundary between the outflows coming from the two cells. Cell motion was to the east-northeast and there was a light east breeze at the ground.

This photo (b) shows a 'front-loader' storm with a well-developed, classic shelf cloud ahead of a solid rain curtain. The leading edge of the outflow is rising and condensing as a lowered rim that is white due to reflected light from bright, clear sky to the east. Above it, the terraced surface is where stable air ahead of the cloud bank is being forced to rise and shape a balance between light inflow winds and the forward moving outflow. The system rolled east-northeast from around noon until it weakened after 7 pm. Except where outflow pushed ahead for brief periods, the system faced a light east breeze at the surface, which may have aided in its slow but continuous forward propagation. At all times the heaviest rain, hail, and lightning occurred ahead of the main precipitation area, with only light rain in the system's wake.

6.47 This is a well-developed shelf cloud illuminated by repeated lightning flashes. At top, the light is from cloud-to-cloud flashes within a thick anvil whose lower surface is covered with crude mamma. In the distance, heavy rain reflects light from a bright discharge, creating a pink glow that silhouettes the thick bank of lower cloud. The shelf has a typical laminar edge and ragged base near the rain. After it passed, the bank grew into a short line of new cells farther south. This view was at the southern end of a southeast moving MCC which produced winds over 100kmh (62mph) and small hail as deep as 20cm (8in) in an area about 50km (30 miles) to the north and northeast.

6.47 (SE New Mexico, 22/5/85, 2 AM MDT, NW) ▶

▲ 6.48a (5:15 PM, SW, 46deg)

6.48b (5:17 PM, SW, 80deg) ▶

(Texas, 22/5/87, 5:15–6:26 PM CDT)

6.48a–e This series of five photos shows the evolution of an unusual roll cloud plus another, similar roll the same day. Isolated storms were embedded in broken mid-cloudiness on the north side of a slow-moving cold front.

A storm producing large hail is moving slowly from the southwest in (a). Surface winds from the northeast, outflow from the north, and an inflow on the southwest side are creating an unusual flow pattern around the system. This has led to the peculiar laminar cloud which is moving left and shifting gradually south. Outflow is being lifted and squeezed along a local zone of converging airstreams, but the process is difficult to explain. The appearance is similar to a shelf (note second rim above) but without the other causes and usual wind patterns. It is possible that this odd cloud began as outflow lifted over the shallow northeast surface flow at the point where this flow was twisting and rising to meet the inflow.

This overview in (b) shows that the long cloud line heads southeast near the rain, then more southwards farther away. Also, there is a lowering near the left end of the rain, part of which is hidden behind the rain.

After only three minutes (c) the cloud is shrinking quickly. The smooth surface is being replaced by small bumps where internal convection is taking advantage of the weakening outflow.

The tube continues to move away in (d). It resembles a roll cloud without the 'roll' on its axis. Another closer look at the higher, main base reveals lowered fragments and a general slant to the right, indicating air motion towards the inflow area. The storm is now experiencing a period of rebuilding but the visible proof (under the updraft) is still hidden by rain.

▲ 6.48c (5:20 PM, SW, 30deg)

▲ 6.48d (5:29 PM, SSW, 62deg)

The roll in the previous photo became thinner, fragmented, then evaporated after a short time. It did not reform and the storm passed by to the north. About 60km (37 miles) to the southwest, we encountered another storm with almost identical structure and similar roll cloud (e). It also moved southwest and evaporated soon after this picture was taken. Similar, unusual cloud features at different places and times are quite rare and offer an opportunity to assess a situation on the basis of few differences rather than many similarities, since the number of unknowns is at a minimum.

6.48e (6:26 PM, WSW, 18deg) ▶

A shelf cloud or gust front will, on rare occasions, become a roll cloud. This is a long, horizontal, narrow, tube-shaped cloud which is completely detached from the cumulonimbus base. Roll clouds also form separately at the storm's rear, or any other place where the cool air is moving gently into warmer, drier air, and can sometimes be seen rolling slowly about their horizontal axes.

No discussion of a storm's outflow features is complete without reference to those bits and pieces of black stuff that form and vanish in the transition from warm to cold or one stage to another. Those cloud fragments, called scud (sometimes called fractus), form in patches of lifted air that were moistened by falling rain or by mixing with the cool, moist outflow. If the lifting is over a larger area, a layer of more orderly cumulus fractus can result but in most situations the effect is quite localised and brief, with the irregular scud evaporating as soon as lifting weakens or the surrounding drier air mixes with it. Scud forms regularly just ahead of the leading edge of the outflow and may thicken, rise, and join the shelf cloud base. It also forms around or under the wall cloud during its development stage, and may appear in clear sky when cool air has advanced beyond the shelf cloud at the ground. The scud can form very near the ground, presenting the apprehensive observer with a frightful, sinister portrayal of mere, random condensation processes.

6.49 A storm to the west has spread a ragged shelf cloud forward overhead. At the leading edge of the cooler outflow, a roll cloud has formed. It is lighter and lower than the shelf, with the usual smooth, cigar shape. Although it straddles higher land on both sides, this is probably a coincidence and unrelated to its cause. A more likely explanation is that the outflow, on passing a more distant hill crest, slid down the lee slope, setting up a rolling motion as it undercut warmer air.

▼ 6.49 (Texas, 23/5/87, 7:56 PM CDT, SE, 34deg)

6.50 A severe squall line is moving east into tropical air. For hours one cell after another formed on the south end and migrated northeast up the line, maintaining continuous flashing in the upper portions of the cloud. The system as a whole also cycled through stronger and weaker phases which lasted about half an hour each. One of these stronger periods is underway in the photo, and is marked by many lightning bolts out of the rearside where powerful updraft bursts hang back over the outflow area. The silhouetted tube is a roll cloud moving slowly out and away from the precipitation along the front of an outflow surge. It is interesting that the outflow can glide smoothly well above the ground but create only ragged scraps of scud in turbulent motions lower down. Several of these rolls were seen that night, each forming, thickening, moving northwest, then evaporating again as the cool air slowed and mixed with the drier air outside the system.

6.50 (S Texas, 29/5/87, 12:45 AM CDT, E, 40deg) ▶

▲ 6.51a (5:52 PM, SW, 180deg)   ▼ 6.51b (5:56 PM, S, 52deg)

(W Oklahoma, 27/6/83, 5:52/5:56 PM CDT)

6.51a,b A severe storm about 20km (12.5 miles) ahead has sent a large pool of cool outflow rushing forward. In (a), the ultra-wide angle captures the appoaching gust front as it fills the view with a smooth, curved rim, below which a second, more ragged lower rim marks the edge of strong winds that are ploughing air upward. This is producing low scud (just right of road) and a wall of dust (lower left), both of which are rendered less obvious by the wide-angle lens. Beyond this in the cool air, a turbulent shelf-like base precedes the rain area at the horizon. The storm was moving east at the north end of a line of activity. See also 6.5lb.

(b) is a close view of the passing gust front (seen distant left in (a) which is accompanied by a wall of dust. The outline of the dust parallels the wind motion: at the leading edge it is farthest forward at the ground; above this it curves up and back into the lowered shelf cloud base; while behind the wind surge it rises slowly in a steady horizontal flow. Windspeeds of 60–70kmh and a sharp temperature drop raced east about 40kmh for about 10km before slowing down again.

◄ 6.52 (Texas, 14/5/86, 5:50 PM CDT, ENE, 62deg)

6.52 A moment ago, the visibility was fine and we had a good view of the dark, low updraft core of a supercell ahead to the east. Then, suddenly, massive clouds of reddish dust were lifted from the dry fields by very strong bursts of descending air (microbursts), probably from the rear flank downdraft. Although the storm passed right over this location minutes earlier, the thirsty ground was only greeted by a handful of large hailstones. This system generated several short-lived tornadoes over the next half-hour before spawning two violent ones later in the day.

MAMMA

The anvil of a cumulonimbus in the light of the setting sun displays the most ornamental of all the visual features of the storm, the mamma. Although mamma are a mere by-product of no consequence to the structure of the storm, and in many respects represent simple shapes and mechanisms, they have eluded simple explanation. Indeed, their shape is too simple! The pendulous breast-like clouds which appear to hang beneath the anvil have usually been described in terms of instability, as if that were sufficient; but no experiments or other phenomena are quoted to indicate how this fascinating shape arises.

One basic feature of the base of the anvil is that it separates the dry, cloud-free air below from the cloudy air above which is far above the condensation level and therefore contains a considerable quantity of water in the form of ice crystals. If this boundary subsides, the air below will be warmed more rapidly than the air above, which will be cooled by the evaporation of the ice. For every 100m (110yd) of descent, the cloud air will become about 0.2 deg C colder than the clear air below it, and will therefore begin to sink into it while the clear air rises into the cloud between the descending mamma.

Although initially the mamma have a smooth outline it would be expected that, like the top of a rising thermal in cumulus, the surface would assume a cauliflower appearance. Sometimes wrinkles and broken shapes do appear, but as a rule they remain very smooth and this can be seen in the bright illumination to which they are subjected. Because the outline remains sharp it is not thought that evaporation of the cloud in the wrinkles is a likely cause of the total disappearance of the mamma.

The other factor present is the large number of falling particles in the anvil which have two important effects as they fall out of the cloud. First, they begin to be evaporated in the dry air under the anvil and, in so doing, cool the air probably to a greater extent than the subsidence already described. The falling particles therefore create a negative buoyancy which gives rise to the same instability. Secondly, they continue to descend out of the colder air into air below which has not been cooled and caused to descend. They are therefore retarded in their fall, causing a crowding of particles at the fallout front, which sharpens the outline and provides a buffer zone which smooths out the cauliflower-like wrinkles.

When the anvil base is steeply inclined to the horizontal the mamma are often elongated up the slope. When the anvil is very thin we may see brighter skylight in the gaps between the mamma where the clear air has ascended into the cloud.

When anvils are viewed from above by satellite an interesting effect is observed in Channel 3. In the early stages of the growth of the cumulonimbus the cloud looks black because the Channel 3 component of sunshine is completely absorbed by the larger particles. But later on, when the storm has passed its maximum intensity, the anvil becomes grey, indicating that we are now viewing smaller particles which scatter sunshine by diffraction. This is thought to be due to the sedimentation of the particles in the anvil, the larger ones falling through the base, and the smallest ones remaining at the top. This indicates that sedimentation is a significant mechanism in the anvil, which therefore deserves the kind of consideration outlined above.

▲ 6.53 (Kansas, 7/5/86, 8:37 PM CDT, E)

6.53 In this almost vertically upward view, the mamma are beginning to evaporate and become transparent, especially at right. The smaller particles which are dragged down in the wake of the fallout front can disappear suddenly, leaving only the larger particles which are more transparent. This causes a blurring or double-image effect such as seen here, and could indicate the presence of a relatively wide range of particle sizes in the anvil.

6.54 Mamma-like bulges can also appear on the bottom of falling precipitation, especially if the particles are small and leave the cloud in large bursts. The droplets are swept down in the downdraft but crowd together at the fallout front as they experience increased air resistance just below the downdraft.

▼ 6.54 (Alberta, 3/5/82, 7:27 PM MDT, SW, 30deg)

Relationship of Mamma to the Storm Environment

There is something about a sky full of dark, hanging bulges that brings to mind images of impending doom but the connection between mamma and unusually bad weather is probably exaggerated. Mamma do not

6.55a,b These show the formation of mamma above an outflow pool.

In (a), a collapsing storm with heavy rain to the right has spilled cold outflow to its east and southeast. At the leading edge is a long, dark line of large cumulus. Between the line and the rain, subsidence has thinned out the clouds and begun early mamma development.

(b) The sky is now full of mamma which have even invaded the thicker cloud bank in the distance. They are large (note the wide view) since the cloud base is relatively low. The outflow depth is shallow and as the cool pool extends farther from its source, sinking air induces mamma to form. They may also be aided by the upward-sloping base trailing away from the cumulus line.

Minutes later, the mamma began to lose their sharp edges and rounded appearance as evaporation took over. Shortly thereafter, an abrupt transformation occurred as the entire sky lost all detail and reverted to a uniform grey altostratus sheet with faint signs of fallout. The outflow had stopped moving forward earlier, and light west winds became strong south again coincident with the sudden change.

▲ 6.55a (11:11 AM, SE, 110deg)

▼ 6.55b (11:27 AM, SE, 110deg)

6.56 Backlighting along the anvil edge of a passing shower reveals large mamma that may be initiated by pockets of descending raindrops.

▼ 6.56 (Wyoming, 12/6/86, 6:50 PM MDT, NNE, 46deg)

guarantee severe weather or necessarily even suggest that any major convection is underway nearby. Since they form best on a uniform part of an anvil, they occur most often at the storm's rear or outflow flank, where the anvil has settled or slowed down in its motion. For the same reason, they appear more often after the storm has passed the mature stage.

There are other important relationships to consider, though, which can tell us much about the behaviour of the atmosphere that day. Lower, thicker anvils lead to less well-developed mamma. Higher anvils that were formed rapidly have many smaller particles which are conducive to sharp, solid pouches. If the air is quite dry or the anvil has little fallout, the mamma will be ragged. If the bulges are uniform in size, the anvil is of a uniform thickness. If the undersurface is sloped, indicating shear, the mamma will also be stretched and may be enhanced by the downward motion. And, of course, the subsidence that started the process may indicate strong new development somewhere in the area. A large storm complex will generate successively higher layers of anvil cloud as it approaches a peak in its intensity. The higher anvil can induce a downward pressure on the lower one when it pushes outward like a wedge, causing the sudden formation of a whole sky of mamma to mark a hidden stage in the convective activity.

Mamma provide us with a beautiful and entertaining by-product of the storm environment. They reveal many things about the processes that operate in the anvils of cumulonimbus high in the troposphere and are quick to show even small changes there. Like the other features and unusual manifestations of the storm environment, they appear in an unlimited array of shapes, sizes, colours, and situations. Mamma are one of the adornments of sky, a feast of visual delights in the midst of threatening overtures.

▲ 6.57 (Kansas, 12/6/83, 7:45 PM MDT, E, 94deg)

6.57 The anvil extending back from a distant storm has a sloped underside which expedites mamma formation, seen here as brighter lumps below the shaded base.

6.58 Sunset has turned the sloped base of a distant left-moving storm into a golden glow. Beyond silhouetted altocumulus patches, a field of mamma litter the base. They are small and ragged, indicating widespread fallout and evaporation.

▼ 6.58 (Kansas, 16/6/85, 8:52 PM CDT, SSE, 52deg)

6.59 (Kansas, 7/5/86, 8:40 PM CDT, SE) ▶

6.59 These mamma are loosely arranged in ridges and show wide gaps with faint blue sky where the upward air motions have thinned the cloud layer by penetrating it. The anvil was old and completely detached from any lower clouds or precipitation. Distant mamma are more ragged, indicating some fallout is still present there.

▲ 6.60 (Texas, 1/6/85, 8:45 PM CDT, NNW, 30deg)

6.60 Silvery mamma are catching the setting sun's rays just below the base of a high, backsheared anvil (edge at lower left). The falling ice particles have formed very detailed pouches that are flattened and striated by the upper winds.

6.61 Mamma extending throughout the anvil and almost to its edge.

▼ 6.61 (Texas, 28/5/87, 5:02 PM CDT, NW, 80deg)

▲ 6.62 (Wyoming, 12/6/86, 9:30 PM MDT, N, 80deg)

6.62 Large mamma, made visible by lightning within the anvil of a storm moving east. The bulges precede a rain area which is seen as the featureless area lower left. The mamma are being absorbed by the advancing fallout as is evident in the more distant ones.

6.63 (Kansas, 16/6/85, 8:35 PM CDT, E, 12deg) ▶

6.63 These classic rounded mamma, seen up close, become visible only where sunlight is able to illuminate them.

6.64 The rearside of a squall line has been transformed from a smooth, uniform anvil to this unusual pattern. Wrinkly mamma appear beyond the parallel grooves (distorted by perspective in wide view) which are across the flow but down the slope of the undersurface. The pattern of grooves and ridges could occur when descending mamma are impeded by a sub-anvil stable layer and begin to slide down the slope instead. The mamma formed to the east and gradually spread throughout the sky to the edge of the anvil (see 6.61). The grooves formed later, then disappeared again long before the mamma did.

▼ 6.64 (Texas, 28/5/87, 4:27 PM CDT, ESE, 110deg)

▲ 6.65 (Colorado, 13/7/76, 7:50 PM MDT, S, 94deg)

6.65 We are on the north side of a weakening thundershower which is moving slowly southeast. The prevailing surface flow is light northerly behind a weak trough. Where this breeze meets outflow pushing north from the storm, several very smooth roll-cloud elements have formed. Although this roll cloud is not a typical tube shape, it is detached from the main storm base and shaped by the same kinds of air motions well away from the system's true inflow region.

(Texas, 29/5/87, 3:41–3:42 PM CDT, NW, 34–40deg)
6.66a,b A piece of scud has formed (a) in moistened air moving north-northwest out of an area of showers. Since the terrain was

relatively flat and sunshine was obscured by a thick anvil, the reason for the smaller wave-like cloud is unknown. It is possible that the low-level moisture was spreading into warmer and drier air in tongues or sheets, and that the smoother cloud was one of these being forced up locally by a thermal. The larger part shows signs of rolling, with the base pushing forward but the top curling back in lighter winds.

One minute later in (b), the cloud has changed but the mystery remains the same. The thin clear strip above the smooth surface could be a thin layer of drier air being wrapped into the cloud. Or, could it be a locally stable patch where air is descending behind or below the scud due to rolling motions?

▼ 6.66a

▼ 6.66b

7

EYES IN SPACE

As exciting as the sky chase can be, there are limits to what we can experience, earthbound as we are. But what we cannot see from the ground is available to us now through our eyes in space – satellite photography. The larger patterns and influences which interact with factors on the visible, local scale are displayed in their rhythmic glory as we look down on the Earth. Combining these new observations with those we make from Earth's surface gives us a much enriched understanding of weather processes.

THE NEW OBSERVATIONS

From a point at sea level on the Earth's surface we can see cirrus clouds which are at a height of 10km (6.2 miles) when they are at a distance of about 130km (80 miles). Clouds much farther away are below the horizon. Weather systems – cyclones, air masses, fronts and anticyclones – usually extend over several hundred or even a few thousand kilometres. To understand their structure early meteorologists made synoptic charts, which are maps of weather observations made at the same time over a large area.

At first these observations had to be collected by post. With the invention of the electric telegraph, and later radio, it became possible to collect them within an hour or so. Nowadays the exchange of this information all over the world is very rapid, so that a weather forecasting office constructs at least four synoptic charts every day, and many more for their own local area.

Most of the information is simple and numerical and relates to temperature, wind, pressure, humidity, and visibility. But a very important part is coded information about clouds which can give only a very much simplified description of the sky. The meteorologist must learn by experience what these messages mean, and this is done by looking at the sky and making his own coded message to describe it. Even more than early meteorologists seamen, who depended on wind power to travel and to avoid or survive the worst parts of a storm, built up a three-dimensional picture in their minds of the structure of a cyclone. This picture has proven to be very good through subsequent close examination of many charts and radio-sonde measurements of wind and temperature in the upper air.

Satellite observations began a quarter of a century ago, and have now been developed into an excellent routine. We describe the NOAA (National Oceanographic and Aeronautics Administration) series of polar orbiting satellites which carry the AVHRR in some detail, because they have provided the best pictures. These have been collected in an archive available for detailed study in the University of Dundee. NOAA has launched eleven satel-lites of this series, and we describe the latest of the series now in operation.

THE ORBITS

The satellites are in orbit around Earth at a height of about 840km (522 miles). One orbit takes about 100 minutes. Their horizon is about 3,376km (2,100 miles) distant from the satellite and about 3,100km (1,926 miles) from the point beneath them along the Earth's surface. The swath observed is just over half the strip that can theoretically be seen; but the part more than 1,500km (932 miles) from the centre line is viewed so obliquely that the information is not generally useful in building up a picture as seen from directly above.

The orbit is so high and the air so sparse that the satellite remains in orbit as if without friction for several years. The equator is 40,000km (24,856 miles) long. There are 14.4 orbits of 100 minutes each in 24 hours. To cover the whole Earth in one day the swath observed would have to be at least 40,000 ÷ 14.4km wide, ie about 2,780km (1,727 miles). Thus the swath 3,000km (1,864 miles) wide gives a small overlap at the equator and much more in higher latitudes.

This is achieved by using the rotation of the Earth and keeping the plane of the orbit fixed in space. If the orbit were to be exactly fixed it would have to pass over the poles; but it is more convenient to have the satellite passing overhead at roughly the same solar time (the same time of day) every day, just as we make all other meteorological observations at the same time each day. Because the Earth goes round the sun once a year the plane of the orbit must also rotate once a year, and this is achieved by making the orbit pass to one side of the pole but, as it happens, to include the pole in the swath. Thus the scene in the neighbourhood of the pole is observed, on average, 14.4 times a day.

The orbit does deviate over the months from the intended orbit because of measurement inaccuracies and uncertain irregularities in the shape of the Earth, and it has to be maintained by continual adjustments made by small jets on the satellite. Since this consumes mass the useful lifetime of a satellite is limited to a few years.

HOW THE PICTURE IS BUILT UP

The camera on the satellite has a mirror which rotates six times a second. On each rotation it observes a strip across the swath which is divided into pixels (picture elements) which are square vertically beneath the satellite. Because the pixels subtend the same angle at the satellite they cover an elongated area at the edges of the swath which

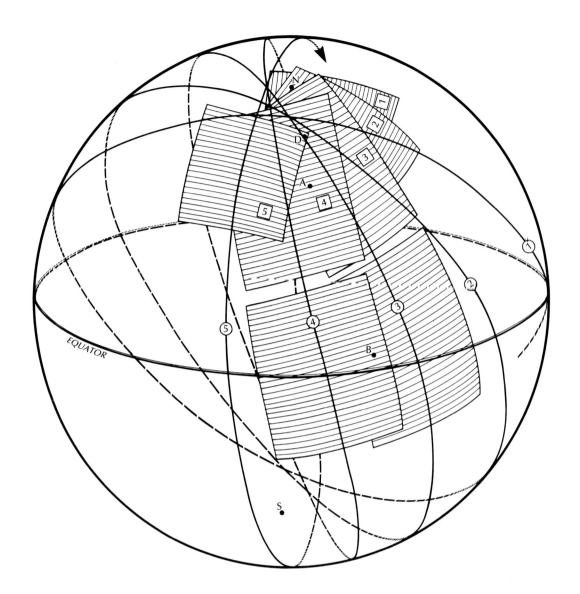

7.1 This shows five successive orbits as they circle the earth. Parts of the swaths of these orbits that are received by stations A (at about latitude 55 deg N) and B (just north of the equator) are shaded.

A can receive parts of some orbits previous to 1 which pass between it and the north pole, which is in every orbit's swath. B can receive only two successive orbits, and equatorial stations

that are near the centre line of the swath can receive a message from only one orbit. The station D will appear in all the pictures received at A as indicated.

The orbit is very close to a circle in space, but its path along the Earth is a complicated coil because the Earth is rotating inside it.

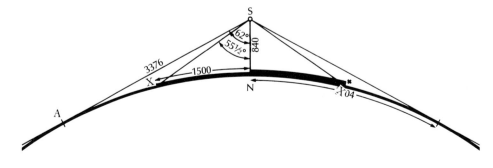

7.2 The size of a cross-section of the swath. The satellite is at a height of 840km (522 miles), and its broadcasts can be received by any station within 3,104km (1,929 miles) of the point N below the satellite S. The straight line distance of A

and B from S is 3,376km (2,100 miles).

The swath (XNX) is 3,000km (1,864 miles) wide and subtends an angle of 2 × 55½ deg at S.

The angle that SX makes with the horizon at X is 21 degrees.

130

is about three times as wide on the ground as directly beneath the satellite.

The satellite travels at 450km (280 miles) per minute at its height of 840km (522 miles), and covers ground at about 400km (250 miles) per minute. The swath is being extended at a rate of 6.7km (4.2 miles) per second, so that with six mirror rotations a second the pixel beneath the camera at N is 1.1 x 1.1km (0.7 x 0.7 miles) squared. At the edges it is 2 x 3km (1.2 x 1.9 miles) squared (at X). The satellite broadcasts its observations as it makes them. To receive this message it is necessary for the satellite to be above the horizon of the receiving station. Thus, a section of the swath about 6,200km (3,853 miles) (2×XN) long can be picked up from a satellite that passes directly overhead. Parts of the neighbouring orbits can also be received, but these are shorter. The nearer to the pole a station is located the greater the number of orbits that may be sampled.

The system is designed to have two satellites in orbit which are at right angles to each other, giving every point on Earth at least four passes a day. In the latitude of the British Isles a station can receive four successive passes of useful length and the station itself will appear on two, and in some cases three, because the overlap is about half the swath width at 60 degrees latitude. Thus the picture is like a continuous strip of TV picture being transmitted at the rate of 6.7km (4.2 miles) of length every second.

WHAT THE PICTURES TELL US

The cameras pick up any radiation coming from the Earth. This is of two kinds. First, there is sunshine which illuminates the scene by day and while some of the energy is absorbed, particularly by dark objects, much is scattered back to space. This consists of radiation with wavelength in the range $0.2–4.0\mu$m when it leaves the sun. Ozone in the stratosphere absorbs anything of wavelength less than 0.3μm. The visible range is from 0.4 to 0.7μm and the radiation is most intense at 0.5μm, which is green. We call the range $0.7–1.5\mu$m the near infra-red, and from the optical point of view it behaves just like the visible and it can be just as easily turned into visible black-and-white pictures. The same goes for the middle infra-red from $1.5–4.0\mu$m.

There is also the radiation emitted by Earth itself. At a temperature of about 300 degrees Kelvin (27 degrees celsius) this consists of wavelengths in the range 3–40 μm. The maximum intensity is for wavelengths close to 10μm. All the radiation with wavelength greater than 15μm is absorbed by the water vapour and CO_2 in the air. The intensity of the emission increases very strongly with temperature so that warm objects are brighter. But we are used to clouds being white and the ground being darker except when it is cold and covered with snow, and so the photos in deep infra-red ($10.5–12.5\mu$m) are printed in photographic negative because the highest, and coldest clouds are the whitest and the warm ground in hot sunshine appears very dark because it is very warm.

The radiation passes into the Advanced Very High Resolution Radiometer (AVHRR) which separates out five bands, or Channels, which send the measured intensity

to the transmitter. The channels are as follows:

Ch1 $0.55 – 0.68\mu$m visible deep yellow/red
Ch2 $0.725 – 1.10\mu$m near infra-red (often called visible)
Ch3 $3.55 – 3.93\mu$m middle infra-red
Ch4 $10.5 – 11.5\mu$m deep infra-red
Ch5 $11.5 – 12.5\mu$m deep infra-red

Channels 1 and 2 produce good pictures by day, but are not sensitive enough to detect anything at night, not even in full moonlight. Channels 4 and 5 are so similar to each other that for our purposes they are the same. They produce similar pictures by day or night because sunshine does not contribute anything in these wavelengths.

Channel 3 is the most interesting in many ways because it is just in the overlap region where sunshine and the Earth each make a small contribution. By day the scattered sunshine is about three times more powerful than the Earth's contribution, even though in sunshine it has about ½% of the intensity of the green, and so we can regard it as a kind of light, with shadows and reflection at water surfaces and we print it as if the radiation were visible. By night it looks very like Channel 4 and is customarily printed in photographic negative.

Channel 3 has several optical properties different from Channels 1 and 2. It is strongly absorbed by water and ice, so that clouds with larger sized droplets or crystals (10μm or more) completely absorb it, and appear black in sunshine. With smaller-sized particles there is some diffraction, and clouds with mainly very tiny droplets (1μm or thereabouts) in the top look quite bright. But the scattering by diffraction is directional and so the brightness depends on the angle of the sun and of the satellite. In this wavelength there is very little scatter by tiny aerosol particles (around 0.1μm) and therefore we do not see haze or smoke in this channel. This also means that there is very little skylight such as we ordinarily see by day; we would not get much through the windows in Channel 3 except by direct sunshine, and shadows are much sharper as a consequence, just as they are on the moon. Finally, because of the large refractive index and absorption in water the reflection of Channel 3 is very strong and thus much more intense than in the 'visible' channels. Sea glint is very bright, relative to the other scatterings.

VISIBILITY PROBLEMS

Another satellite was put into orbit for the benefit of oceanographers, and it happens to have done a very good job for meteorologists also. It was called the Coastal Zone Colour Scanner (CZCS). It was switched on only for chosen experiments, and its operation ended in August 1986. On those occasions when it was transmitting within sight of Dundee its message was recorded and it has provided some very good cloud pictures.

There were five useful channels designed to observe the colour of the sea, which varies according to its biological and particulate content. The channels were:

CZ1 $.433 – .453\mu$m   visible blue
CZ2 $.510 – .530\mu$m   visible green

CZ3 .540 – .560μm   visible yellow
CZ4 .660 – .680μm   visible orange/red
CZ5 .700 – .800μm   very near infra-red

To obtain good pictures of the sea colour the camera was tilted about 20 degrees towards the pole so that sun glint was not observed. Since the sea is usually rather dark the camera aperture for CZ1–4 was opened wider than for CZ5, whose purpose was to record the clouds and coastline at the same time. Consequently the pictures were grossly overexposed for clouds, and even dense haze saturates CZ2 and CZ4. CZ1 and CZ3 are very useful for observing haze, particularly CZ1, because the land is much less bright than in CZ3.

CZ5 produced some very nice cloud pictures because of the absence of glint and also because the pixel size was 0.8 x 0.8km sq (0.5 x 0.5 miles sq), which gave some useful detail occasionally.

The pictures of this satellite demonstrate one of the difficulties with channels in the visible wavelength range, namely that too much haze is picked up for the pictures to be good with clouds. That is why the most popular 'visible' channels are near infra-red, which scarcely records haze. On the other hand CZ1, the blue channel, has shown the great value of pictures in this wavelength for the study of very important air pollution incidents.

In the early days it was thought that colour pictures would be more useful than black and white, as they undoubtedly are in observing clouds from the ground. The attempts were not helpful to science because of the intrusion of reduced visibility. If we could have ordinary cameras which would take pictures using narrow wavebands, or even the fairly broad bands of the AVHRR, we could discover much more about the physical properties of the cloud particles than we can with ordinary colour or black-and-white film.

AIRMASSES

As soon as we look at the Earth with eyes in space we see that there are quite large areas, from 1,000 to 6,000km (about 600 to 4,000 miles) wide, in which the cloud pattern is the same all over. There are also many lines of more dense cloud between these regions. We call these regions airmasses. Their characteristics are the result of having been in one area, such as part of an ocean or continent, or over polar ice, for many days or weeks. We may be looking at an airmass in its place of origin which has made it cold or warm, moist or dry, deep or shallow. Or it may have been moved to a different area where we see the cloud patterns associated with its modification into a different airmass.

The most obvious cases of which we now have a much clearer picture are over the wide oceans. There, conditions do not change much from day to night and from week to week, because the ocean temperature changes very slowly.

A cold and always rather dry (in absolute terms) polar airmass which is moved into an area of open sea warm enough to promote convection in it is filled with cumulus clouds which warm it and evaporate water into it. Such cold airmasses often subside rapidly as they move from their source region, and this produces a stable layer which acts as a ceiling, setting a limit to the cloud height. The clouds then become arranged in streets along the wind. The subsidence can often be seen (3.1, 3.11) in the sideways spreading of the streets, and the fact that the air is being moistened is shown by the large gaps between the clouds, for without evaporation they would spread out at the ceiling and form an almost continuous layer.

Without subsidence the cold air would undergo strong convection, and the clouds would grow into tall cumulonimbus with anvils carried away at their tops by the thermal wind (2.12). These make a complete contrast with the low stratus or sea fog which fills a moist tropical airmass as it moves to cooler sea, being cooled by radiation down to the sea temperature (2.8).

7.3 Sahara dust over Iberia. By comparing this picture with 2.13 we can see how effective the tilting of the camera northwards has been in eliminating glint. In 2.13 the glint was in the east, but this is three hours later and so the glint would be near the middle of the pass. The Coastal Zone Colour Scanner (CZCS) picks out the dust very well over the darkened sea, but does not improve on Channel 2 over land, both channels being in the very near infra-red.

7.4 Looking at dust through a blue filter. The CZCS Channel 1 (blue) is very sensitive to dust in sunshine, so sensitive in fact that it obscures a great deal of the land, and it cannot see through the haze at all over Portugal. Furthermore, the clouds are so bright that all detail of their structure is lost. It does detect the condensation trails near the French coast.

7.5 Ocean convective storms. The storms of 2.12 off the west coast of Ireland are seen here in infra-red (Channel 4) and enlarged. The sea is warmer (darker) than the land and the cumulonimbus are larger over the sea even two hours after midday, the warmest time on land.

The enormous anvil clouds show an anticyclonic curvature, caused by the outflow from the storm tops.

7.6 Ocean convective storms. The same storms seen by Channel 3 have 'black' tops because the cloud particles are large, most being ice, which absorb this radiation. The only bright clouds are the small fragments around the bases. The sea is brighter because it is roughened by the wind and the satellite is picking up scattered reflections from the waves.

▲ 7.3 (21/8/80, 11:40 AM, CZ5)

▲ 7.4 (21/8/80, 11:40 AM, CZ1)

▼ 7.5 (15/12/86, 14:03 PM, 4)

▼ 7.6 (15/12/86, 14:03 PM, 3)

FRONTS

The cloudy boundaries between airmasses are called fronts, and we call them cold or warm according to whether it is the cold or warm airmass that is advancing.

There are many different possibilities for the air's behaviour at fronts. The stereotypes are as follows. At warm fronts the warm air is rising up the wedge of cold air and is producing vast sheets of cloud, with cirrostratus at the top of the wedge and arriving overhead first. This gradually thickens and soon produces rain or snow, with the altostratus becoming nimbostratus. On the arrival of the warm air at the surface the temperature rises and there is an increase in the low cloud, while the upper cloud begins to thin out or disappear. At a cold front the advancing cold air shovels up the warm moist air to form dense cloud with usually much heavier rain than at the warm front, but of shorter duration. These active fronts gradually lift the warm air off the surface and close up the warm sector of the cyclone that grows at this point on the front.

Although this picture represents a great advance in meteorological thinking based on the study of charts and radio-sonde measurements, and continues to be taught to young meteorologists as the Norwegian frontal theory of cyclones because it is a very great aid to understanding, it must be qualified by saying that since either airmass may ascend or subside strongly or weakly, or do neither, there are at least nine different kinds of front, even without reference to the intensity of the ascent or subsidence. So each front has to be looked at carefully to see what kind of weather it will bring – heavy rain or snow, a slow or rapid clearance, a sharp or trivial change in temperature, a large or small veer of wind following upon a smooth breeze or gusty gales, and so on. All these features change in intensity as the cyclone develops or declines, moves or stays put.

This is not the place for a text on the vicissitudes of weather forecasting. It is an occasion to enjoy the view; and how much more enjoyable it is than the scenes we find on other planets; and how wonderful is the variety, not only from one place on Earth to another, but in the same place from time to time!

CYCLONES

A cyclone is a vortex of air rotating in the same direction as Earth, but with greater angular velocity. This is the consequence of convergence of air into the bottom of regions where the air is rising and generating cloud and fallout.

This spin-up mechanism is a very fundamental one widely exploited in gymnastics, high diving, skating, and even in using the movement of our arms and bodies to remain upright on our very small base of only two feet. For example, the diver can arrange to enter the water head first by controlling the body's speed of rotation. He does this by drawing in his limbs to increase rotation, or stretching them out to slow it down. To operate this mechanism at all there must be some initial rotation, and in the atmosphere this is provided by the Earth.

The spin-up causes a low pressure region in the centre, to provide a force to counteract the centrifugal force and keep the air moving on a curved path. In a tropical cyclone this is a very important factor, because the drop of pressure expands the air and makes it colder, so it extracts more heat and moisture from the sea, and this sustains the storm for several days. The same is true to a lesser extent in the deepest of frontal cyclones of temperate latitudes; but in them the lifting of warm air by the wedge of cold air is absolutely basic. When the warm air has been lifted off the surface, and the cold front has caught up with the warm front to form an occlusion (2.11), the warming of the cold air by convection may be a major influence in keeping the cyclone going (2.12), or even in creating a new polar air cyclone within the polar airmass.

Occlusions are sometimes wound up like a swiss roll in the centre of an old low, especially when the centre has been almost stationary for a day or two. Other lows have an agglomeration of large cumulonimbus in the middle. In all cases the pattern seems to have its own main features, and from our eyes in space we can see how only one or two observations from occasional ships may not be able to give a true picture of what is going on in the cyclone.

Small cyclones form in great numbers in high latitudes when cold airmasses are over warm sea, but they are of short duration. Again, they can be studied only in satellite pictures at present, because of the wide spacing of the observations made in the arctic seas.

The smallest of cyclonic storms with rain or hail is the tornado. Although they are produced by the same spin-up mechanism in principle they are in one respect very different. They are the by-product of a bigger storm, and their vertical extent is much greater than their horizontal width. Nor are they normally visible from space. We have discussed them in the second half of Chapter 6.

7.7 (21/1/84, 16:14/14:32 PM, 4) ▶

7.7 An active front. A cold airmass is entering the North Atlantic from northern Canada and is full of showers. A cyclone has formed on the front which divides this from an airmass of southwestern origin, much warmer and more stable. This second airmass is streaming northwards into the Norwegian Sea.

The front lying through SW England and NE Scotland is a shelf front, in which the upper air moves forward more quickly lev leaving a shelf of lower level warm air behind to the west of Ireland. This scene is full of interesting cloud types which we could fill a book describing. Nevertheless the beauty and dynamism of its detail are yours without any technical description. The lines, the curves, the fibrous streaks, the dapples, all possess a natural balance, a material ecology thrusting the mild tropical air into the arctic seas.

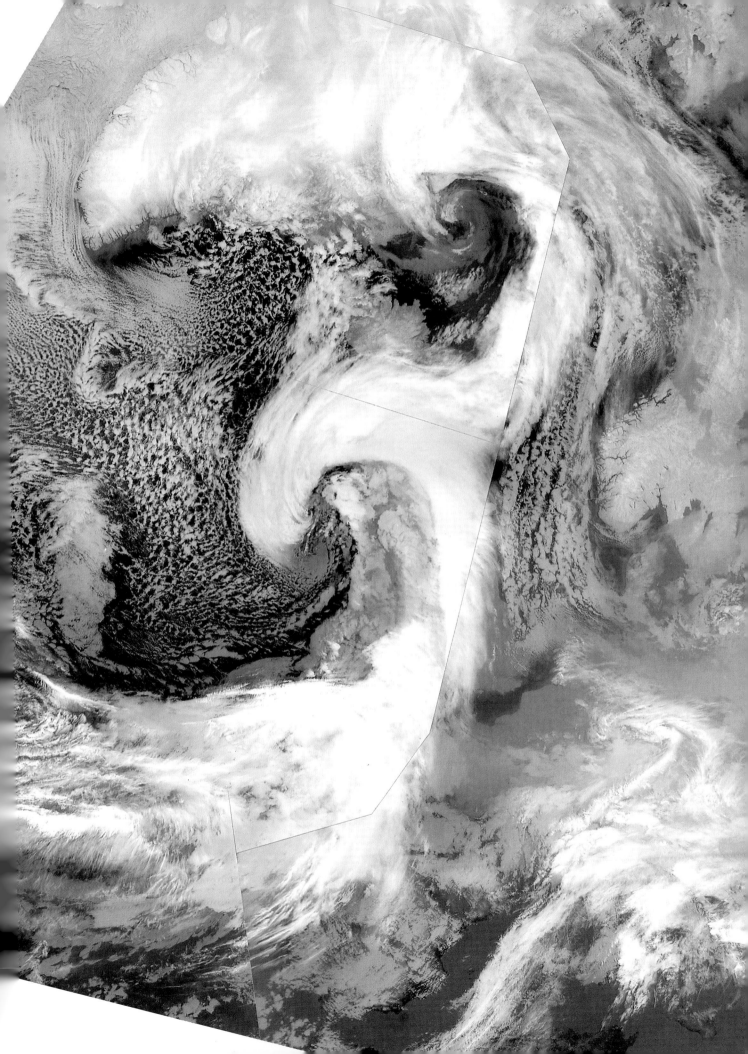

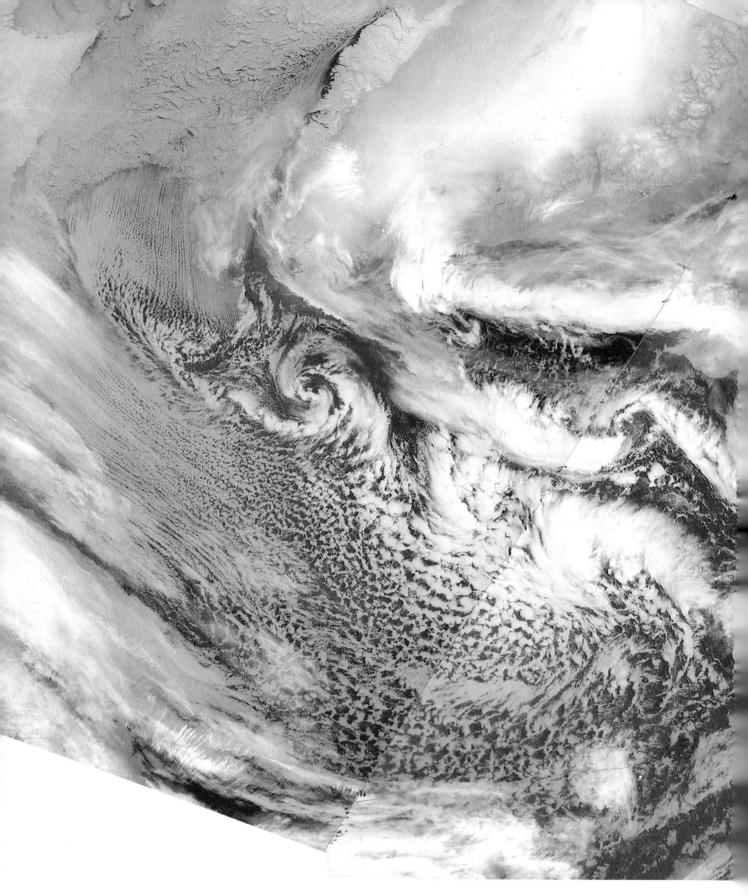

▲ 7.8 (29/1/83, 16:44/15:00 PM, 4)

7.8 A row of upside-down cyclones. This panorama extends from Labrador in the west to Finland in the east. A series of cyclones (not the ones seen here) fed warm air from the southwest into the Norwegian and Greenland Seas and filling the region around Iceland. Then in the next few days there was a surge of cold air from northern Canada and the frozen north around Baffin Island. This spread across to Europe making a front from the tip of Greenland to the Gulf of Finland, with warm air to the north and cold air to the south – a complete reversal of what might be thought normal.

Almost suddenly a row of five cyclones formed at the same

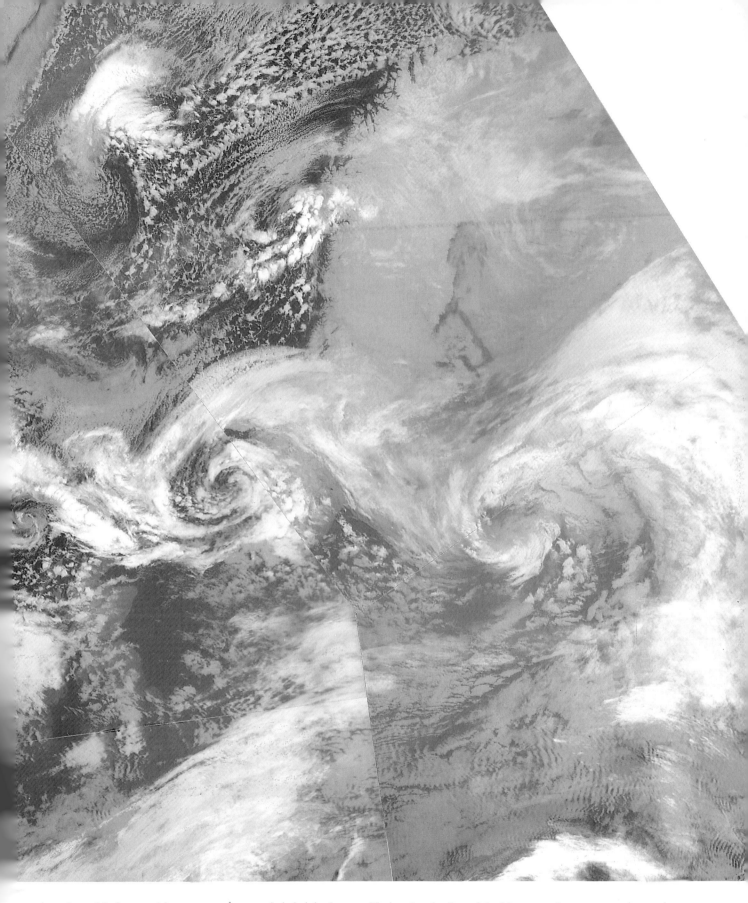

time along this front and here we see them at their height (or depth), for tomorrow they will all be gone. But for the eyes in the sky no one would ever have known about them. Such happenings are quite common, we now know, and at last we can make sense of the occasional observation from a lonely ship reporting winds which contradict what the chart implies.

Notice also the line of the big convective storms to the south of the wavy front with their anvils spreading northwards, and the cells, some closed, some open, to the south of that line.

There is a jetstream curving over northern Italy, with another across NW Europe leading into the end cyclone over Finland.

ANTICYCLONES

When air subsides it spreads out horizontally, and so its rotation is reduced to become less than that of the Earth. To us, therefore, standing on Earth, the rotation appears to be in the opposite direction, and we call the result anticyclones. There is clearly a limit to anticyclonic rotation because it is still in the same direction in space as the Earth's, but less.

By contrast with the violence at the centre of a cyclonic vortex the middle of an anticyclone is a region of calm. Nevertheless the wind can be strong and steady for long periods at the surface although the winds aloft may be much lighter. Jetstreams are not coiled up in the middle, but pushed to the edges. If a front moves into the middle, or if an anticyclone is formed on top of a front, the subsidence causes the evaporation of most of the upper cloud.

At the surface things may be very different, because the subsidence produces a very stable layer, typically at around 1km (0.62 mile) above the surface. This may be at the top of a layer of cloud which becomes trapped. Even the warming due to the small amount of subsidence at this level may be counteracted by the cooling to space through the clear sky above, especially in winter. Over a continent this can produce the unglamorous condition of anticyclonic gloom, which is particularly pernicious when the stable layer is at about the level of the mountains enveloping a well-populated valley in all its air pollution. The air becomes progressively more filthy every day until it reaches an equilibrium, with the dirty deposition of the smoke equalling the new production.

Where there are no surrounding mountains, the calm of a col region where there is no wind may be a worse situation in winter, for the solar heating of the ground may not be enough to make up the radiation losses from the fog or cloud into which the pollution is emitted.

An anticyclone which remains fixed in position for a week or more and appears to push fronts and their associated jetstreams aside when they advance towards it is called a blocking anticyclone. When a big cyclone becomes stationary for several days and all other features are made to circulate round it, it is not called a blocking cyclone, because it does not look like one. The low often sucks smaller disturbances into itself. The necessary concommitants of inflow at the top of an anticyclone or outflow at the top of a cyclone are not so readily seen in the clouds, nor in charts of the conditions at the surface, so the use of the term 'blocking' is descriptive and superficial rather than real. The location of the blocking system in a particluar geographical position is determined by geographical features, most probably mountains. Thus the Iceland low is so called because it is close to Iceland, although it is placed there by the mountains of Greenland. A blocking high over Western Europe may be there because of the presence of the Rocky Mountains or the position of the eastern North American seaboard, or even by the position of Greenland, but a considerable degree of sophistication is needed to see it as such.

7.9 A great polar air cyclone. This storm occupies the North Atlantic from Greenland to Spain, from the Azores to Norway. A polar airmass occupies its centre where convective storms continue to keep the rotation going, some even having their own small whirl. The tip of Greenland is marked by a prominent wave cloud extending down its southeast coast to Cape Farewell, the southernmost point. The cells are open and showery, some with obvious anvils.

Not coiled up in the centre but draped around the southern extremities of this cold airmass lies a jetstream marked by high, white cloud. The southeast section is the most intense, from NW Spain to Austria. What a gorgeous whirlpool of air! The small picture, by visible sunshine, shows where the land lies in this infra-red picture.

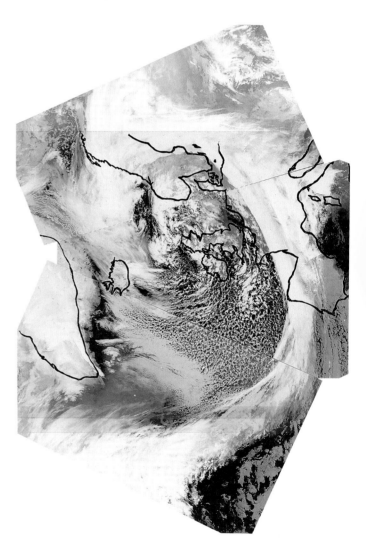

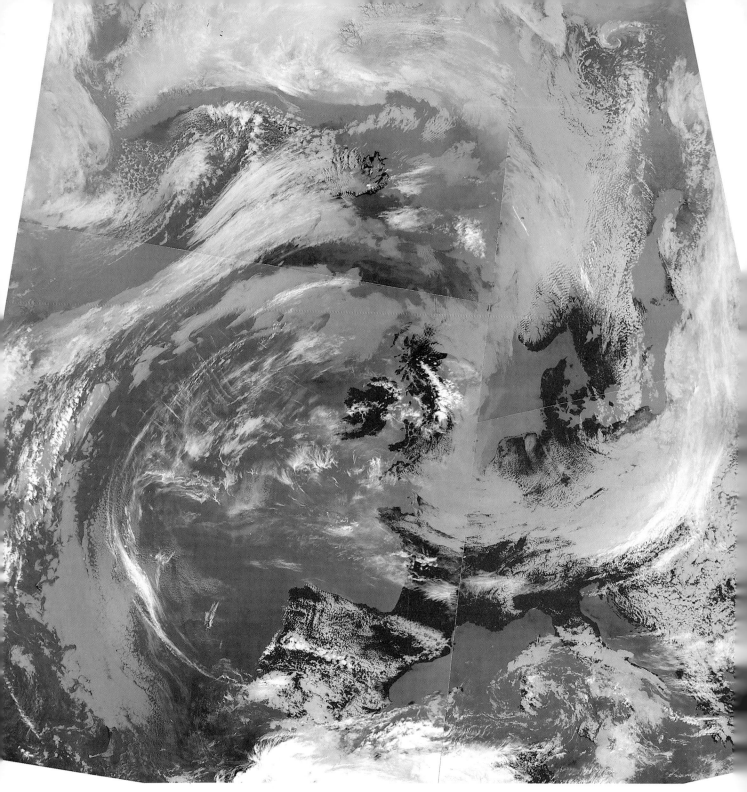

▲ 7.10 (26/4/82, 15:10/13:29, 4)

7.10 This superb anticyclone has jetstreams all round it, the most intense being from the Azores to Iceland round the western side. This continues down Britain's east coast and is then renewed in strength from Central Europe into a low near Leningrad, which feeds cool northern air into Europe.

Iceland sparks off a brief trail of cirrus from its volcanic peaks; Norway sets up lee waves with a tiny streak of cirrus, and protects Denmark and part of Sweden from the cloud by its mountains. The moister air of the North Sea contains typical stratus or fog, while western Britain and Ireland have dry air in the middle of the high.

It is called a warm high because the high pressure is not due to cold air near the surface, but cold air in the stratosphere. The

jetstream circulates anticyclonically around it, and this does not happen with a cold high.

A feeble Mediterranean low is over Sardinia, but the collection of convective storms over Morocco is producing a more convincing example. An old front residue circles round from south Portugal and further north its moist high level air supports condensation trails of the traffic from Europe.

This high lingered in this neighbourhood from 8 April to beyond the end of the month, shunting lows around its northern border.

In its centre are some beautiful feathery patches of cirrus. The land is black because it is warmer than the sea in this infra-red picture.

8
SONIC AND GRAVITY WAVES

This is about two different kinds of wave which drive motion in the atmosphere. What happens when air is heated? It gets hotter, of course, but it also becomes lighter, less dense. Yet no material disappears, so how does this happen? The answer to this question will tell us how heating produces its results, and we have to understand this matter correctly if we are to be sure we have the right explanation for the behaviour of clouds.

THE SONIC WAVE

When heat is applied in a concentrated manner to a body of air its pressure is increased because the velocity of the molecules is increased. The pressure in the surrounding air is unaltered and the heated air expands until its pressure is the same. It pushes the surrounding air outwards.

In an extreme case, when the heat is applied very quickly, we call it an explosion. Some fireworks are designed to make a big noise by creating gas suddenly out of solid explosive materials, and a shock wave travels outwards in all directions as a bang! The bang is a sudden outward displacement of air through which the shock wave travels. After it has passed the air is found to have been moved away from the explosion.

The first atom bomb, exploded in 1945, created an expansion of about 1 cubic km (0.24 cubic mile). This sent out a hemispherical shock wave which produced, at a distance of 10km (6.2 miles) from the explosion, an outward displacement of about 1.6m (1.75yd) in a small fraction of a second – enough to topple a wall or break any window. Such damage was due to the sudden nature of the explosion caused by the extreme heat of the bomb.

If the same total expansion had taken place over ten minutes the outward displacement of 1.6m (1.75yd) at a distance of 10km (6.2 miles) would also have taken ten minutes, but the velocity of 1.6m (1.75yd) in ten minutes could not have been detected by the most sensitive anemometer in the world among all the other movements of air that continually surround us.

Nevertheless this important outward displacement would have occurred, and the heat applied over ten minutes would have produced as big a convection current into the stratosphere as was observed when the atom bomb was exploded.

The important point is that the pressure wave travels outwards through the air with the speed of sound, which is about 316m/sec (1,100ft/sec), or around 1,100kmh (684mph). It travels at this speed for small or big, sudden or gradual expansions. The outward displacement becomes unobservable at a quite short distance – less than 10km (6.2 miles), even for a very big expansion of 1 cubic km (0.24 cubic mile).

We can therefore say that the heating and cooling of air, by whatever cause, can be thought of as instantaneously decreasing the density of the air in question; and the movements which achieve this result travel away into the wide atmosphere and are virtually impossible to detect, having no observable effects as they pass at the speed of sound.

These pressure waves are not dispersive to begin with, which means that they are not steadily spread over a longer time of passage. When we throw a brick into a lake the water level must rise, and the waves which spread out as rings from the splash have the effect of very slightly raising the water level as they pass. But water waves are dispersive because they are gravity waves: instead of one single jump of water level the rise takes a long time and is only finally achieved after the passage of several ups and downs.

Why did we choose the radius of 10km (6.2 miles)? Because beyond 10km (6.2 miles) the wave would be approaching the top of the atmosphere. About three-quarters of the atmosphere is below that height. What happens next? At sea level the wave is expanding like a circular wall around the centre, but it loses its importance in the very rarified air upwards, and by the time it has gone 100km (62 miles) the wave is like a vertical wall reaching to the top of the atmosphere. At that stage it has become like the waves on the surface of a lake, and it is what we call an external gravity wave.

It is still a wave transmitted by the air pressure, but the pressure is maintained by gravity acting throughout the depth of the atmosphere at the wave front. It so happens that the speed of these external gravity waves is a sort of average of the speed of sound through the depth of the air, but since that speed is not the same at all heights and the kind of average varies with wavelength, the speed also varies a little with wavelength. The consequence is that the waves are dispersed, but not so widely as the waves on a water surface. A single pulse is dispersed into a series of about seven oscillations at a distance of 5,000km (3,107 miles) from the explosion. The oscillations are smoothed over if the expansion takes several minutes, because the oscillations have a period of about two minutes in the case cited, and take only about fifteen minutes to pass by. At the antipodes of the explosion the pulse would have spread further, taken about an hour to pass, and would have had many more oscillations. In the famous case of the explosion of the volcano Krakatoa in 1883 the wave was barely detectable after passing to the antipodes and back to Java, where only the first and largest oscillation could still be detected.

▲ 8.1a

▲ 8.1b

8.1 A stone's throw. (a) The single initial pulse when a stone is thrown into water is accompanied only by small disturbances due to splashes, except at the centre.

(b) After a few seconds the single pulse has spread out into a succession of twelve or more wave crests. This is dispersion, caused by the greater speed of the longer wavelength components.

All this means that, for example, the daytime heating over Spain or North America causes an outward-moving pressure wave which displaces the air outwards across the coast. This lowers the density of the air over the warmed land. The waves, travelling at 1,100kmh (684mph), are mostly well out to sea by midday. As they pass they do not have any observable effect and the change in density is confined to the originally heated air.

THE CONVECTION WAVE

Having got the air heated and its density reduced with no apparent motion, the convection starts. Cumulus clouds, thunderstorms, sea breeze fronts, and all the meteorological phenomena which result from hot air rising may be seen. But these do not directly involve wave motion, so what is that second kind of wave – the slow-moving ones produced by the force of gravity – which insists that a stratified atmosphere should return towards its equilibrium position when disturbed from it?

INTERNAL GRAVITY WAVES

Most of the time, in most places, most of the air is stably stratified, so it is capable of transmitting internal gravity waves. These are distinct from external waves which depend on the difference of density between the atmosphere and outer space or, in the case of water waves, on the difference in density between the water and the air above. Internal waves depend on the density differences between adjacent layers within the air. They travel much more slowly, at speeds comparable with typical wind speeds. Because they are very dispersive they are easily observable only when one wavelength has been picked out specially. In the case of lee waves (see Chapter 9) that one wavelength is the one which travels through the air with the speed of the wind but in the opposite direction, so that it becomes a train of standing waves, very often with stationary wave clouds in the wave crests.

When strong convection develops over land a wave causing the air to sink spreads out into the neighbouring sea and evaporates the cloud there. This is not a simple process because it sometimes produces what looks like a single fairly concentrated downward pulse. At the same time a sea breeze front moves inland and the phenomenon has the appearance of an inverted version of that, moving out to sea.

A most unusual form of pulse has been studied in Australia. The sea breeze on the east coast of Queensland seems to surmount the mountains of the York Peninsula, and as it descends the western slopes it encounters the breeze from the western coast. It then sends a pulse westwards through the air which, surprisingly, has remarkable non-dispersive properties and travels many hundreds of kilometres towards Tennant's Creek in the Northern Territory. There it is called the Morning Glory. Other cases have been seen in different parts of Australia. It seems unlikely that a similar phenomenon will not soon be found elsewhere.

REFLECTED MOUNTAIN WAVES

Light and electromagnetic waves are not the only waves in the atmosphere which undergo reflection. Like water waves reflected at a bank, or sound echoed off a wall, the physical barrier of cliffs and mountains reflects pulses when a new airmass is impeded by them.

The reflected waves are visible in the clouds when the structure of the airmass is such as to trap the waves like lee waves. This may be very common because many lee waves are trapped by the wind profile in which the velocity increases with height. But the problem arises in the stage before the lee wave train has become set up as standing waves. Thus a cold shallow airmass flowing up against a barrier which it is not going to flow over the top of will be subject to oscillations which will send a pulse back upwind. The duration of the waves may be short because they will travel away and be dispersed as the flow settles down. Their existence is therefore less likely to be seen on satellite pictures than the standing waves which remain in the same place for many hours.

While good examples are rare they are not difficult to recognise (see 8.2), and they usually consist of rings of waves appearing to radiate from the reflecting barriers. Cases have been observed in the Bay of Biscay due to reflections from the cliffs of Brittany when air has been streaming from the Atlantic.

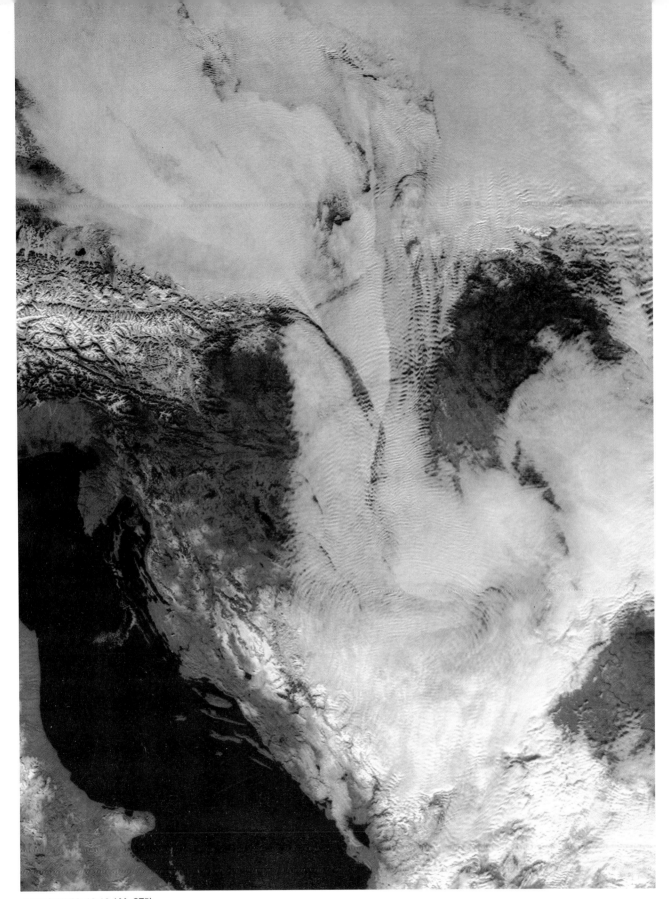

▲ 8.2 (23/1/83, 10:16 AM, CZ5)

8.2 Mountain wave reflection. A shallow layer of cool air containing thin stratus is penetrating into the plain of Hungary and northern Yugoslavia, and quite ordinary lee waves are apparent. But in southern Yugoslavia where the airstream meets higher mountains we see reflected waves which appear to be circular and travelling outwards and northwards. Ordinary lee waves are steady standing waves and relatively easy to deal with theoretically, but these are travelling over the ground and the

mechanism of their production is not in the least obvious. For example, there is no analogous phenomenon in a flowing river except surface tension (capillary) waves, but they have quite different dispersion properties.

Nevertheless many examples of travelling waves have been recorded on anemograms and in time-lapse video pictures from stationary satellites. Because they are quickly dispersed they are seldom observed for more than a few minutes.

143

▲ 8.3

8.3 The Morning Glory. The fanciful name of this phenomenon conveys the feeling of elation which its passage arouses, but it does not begin to describe what it actually is. The nearest to it is the tidal bore on a river, which sometimes comes as a single rather sudden rise in level, but on other occasions it is followed by a series of waves of decreasing size.

So far it has only been reported in Australia, although in several different places there. It is particularly interesting because it appears to travel large distances without much change in form: it is not dispersed like gravity waves.

8.4 Morning Glory. As the name implies, this is an early morning phenomenon, here seen at Burketown in North Queensland at 0630 on 12.10.80. It has been suggested that the original pulse is due to the sea breeze front of the east coast crossing the mountain during the night and, after descending the western slopes, it possibly meets the breeze from the west-northwest on the western side of the Cape York peninsula. It has been suggested that sometimes it is not destroyed by heating of the ground by sunshine, which is the usual fate, but may travel even 1,600km (1,000 miles) inland and be seen the next day near the Centre.

There are conditions peculiar to Australia which make it difficult to find another place on Earth for Morning Glory to appear.

8.5 Coast breeze of S India. Sea breezes caused by the heating of the land by day have been known for many years both in folklore and in scientific studies to travel inland as a front. The quality of the air changes suddenly as it passes, and often there is a narrow line of cloud which glider pilots have used to soar for many miles. This, however, was the first spectacular observation of the complementary, upside-down, phenomenon travelling out to sea, a kind of bore travelling under a stable layer, causing the air to sink and evaporate the cloud. The same clearance out to sea has been observed on other coasts throughout the tropics. This extreme case, with the front on both sides of the peninsula is rare, and is most likely in quiet weather in March or April. This picture was taken by astronauts Conrad and Gordon.

144

▲ 8.5 (14/9/66, 07:30 AM) (S–66. 54674 Gemini XI)

◄ 8.4

9
MOUNTAIN WAVES

GRAVITY WAVES

The air is normally stably stratified at all heights above cloud base, which means that a body of air displaced upwards or downwards and then released will tend to return to its original level. This is not true of air below the base of low clouds when convection is taking place and cumulus is being formed. Nor is it true for air inside clouds where air is rising and receiving extra heat from the condensation of cloud. But at night, and at many places during the day, even the air below cloud base may be stably stratified.

Above cloud base only part of the air is cloudy, and clouds usually occupy less than half the sky. Even when there is a complete layer there are other layers with no cloud at all which are stable. So most of the time in most places there are stable layers, which means that if the air is made to ascend in passing over a hill it will tend to fall back to its original level again. But it does not return the simplest way and stop there; it overshoots downwards and oscillates a few times before finally settling at one level in equilibrium. These oscillations are gravity waves.

WAVE DISPERSION

Gravity waves are usually dispersive in a fluid medium. We gave the example of the rings of waves spreading out from the splash when a brick is thrown into a calm lake in The Sonic Wave, p 141. In that case the number of rings increases, not primarily by new ones starting at the centre, but by new ones appearing among those already some distance out. The dispersion occurs because long waves travel faster than short ones, and this is shown by the greater spacing of the outer rings.

Sound waves, which are pressure waves of very small amplitude, are non-dispersive. Indeed, it is very convenient to the listener that the bass notes from an orchestra do not arrive at the back of the hall sooner than those of the piccolo, even though they may have ten times the wavelength. When someone far away is hammering a stake into the ground we hear the separate knocks of the hammer just as we would if it were nearby; but if three stones were thrown into the lake at the same point at five-second intervals, we would not be able to separate out their waves fifteen seconds later. Indeed we could not easily tell how many stones had been thrown in, while we can easily count the separate knocks of the hammer.

In the atmosphere the situation is much more complicated because the wave energy can often travel upwards as well as horizontally. On a water surface all the stability is concentrated in the top surface, and the waves can only propagate in that surface. Any disturbance in the atmos-

phere, even the ascent of a small cumulus cloud, creates waves, and pileus clouds are evidence of that. But only a very small fraction of the energy released when warm air rises goes into wave-making. The theory of the very complex motions in convection currents is so difficult that accurate calculations are impossible. But it has been fairly easily proved that the wave energy is negligible. Even when a cumulus or cumulonimbus arrives at a stable layer in its ascent and can be seen to produce waves, these waves are quickly dispersed, not as flat rings as on a water surface, but upwards and downwards through the air as well; and very soon they cannot be detected.

The atmosphere is full of these very weak waves travelling in a multitude of different directions and dying out without having any significant effect on the clouds or weather. Even so, there are a few very special situations which produce quite spectacular wave phenomena, and we now look at these.

LEE WAVES

Lee waves are waves which stand still in the lee of an obstacle or other disturbance, most usually a mountain. Other disturbances seldom produce anything comparable with mountain lee waves.

When we see lee wave clouds motionless in the wind the air ascends into the upwind edge as it passes through the condensation level. It descends through the condensation level at the downwind edge where the cloud evaporates. The wave is a standing wave. It travels through the air at the speed of the wind, but in the opposite direction. This is a quite subtle phenomenon because the wind may have different speeds, and even different directions, at different heights. Normally there is only one wavelength which has the right speed to do this, but sometimes there are two, though this is rare. On a very significant proportion of occasions there may be no waves fulfilling the required conditions.

When standing waves are possible it can be quite a complex problem to calculate the wavelength. Nature has no difficulty in sorting the matter out very quickly because any other waves than standing waves simply travel away, while the standing waves grow in situ up to the size determined by the complex combination of circumstances, which includes the precise shape of the mountains as well as the temperature, wind direction, and speed at all heights.

Nevertheless it is fairly easy to estimate crudely, from wind and temperature soundings, whether there is likely to be standing wave motion in a particular place. A forecaster at an airbase can give very helpful advice to an aviator on where to look out for waves. The wavelength, amplitude and position are quite another matter, and observational experience is very valuable in any particular case.

Aviators with passengers prefer to avoid waves because they alter the height of the aircraft and often contain patches of turbulence. Glider pilots, on the other hand, find the free lift available in waves quite delightful.

The fundamental principle of lee waves is that they must be trapped within a layer low down in the airstream so that their energy is not lost upwards. This is most commonly achieved by having a stable layer somewhere low down in the atmosphere with lesser stability and stronger winds (in the same direction) higher up. Even so, many airstreams without such prominent features do sometimes trap waves in the lower layers.

The best waves from the visual point of view are those produced when a layer of low cloud is thin or not quite complete. Definite gaps are then produced in the troughs of the waves where the air has been caused to descend below its condensation level and the wave crests contain characteristically smooth arch-topped wave clouds.

The wave amplitude is the difference in height achieved by a particle travelling with the wind through the waves. This is greatest at the top of a cloud layer especially when it is the only layer. The radiation from the cloud top cools the cloud and, as a consequence, also cools the air below the cloud. This produces a strong inversion (very stable layer) at the cloud top because the air above is not significantly cooled.

When the cloud is a complete cover it may not be possible to see the waves from below, but our eyes in space get a very good view. They tell us that ship-wave patterns are very common. That pattern moves along with the ship and so, for the ship, it is a standing wave pattern. A duck swimming on a calm pond makes a similar pattern. An island in isolation in the ocean may be hidden by a cloud layer, but its presence may be proclaimed from the cloud tops for this is where the amplitude is greatest (2.1).

One very surprising feature of lee waves is that the patterns are always much less complicated than the mountain shapes which produce them. Airstreams over Ireland and Scotland show this very well. Unlike waves on water and sound waves, the waves from different mountains are not simply added together as if each mountain had made its own pattern independently of the others. If that were to happen a second mountain might double the amplitude of the waves made by the first. Or, if its phase were different, it might cancel them out or change the phase without altering the amplitude. In practice it appears that the air holds very tenaciously to the phase and amplitude set up by the first mountain. Also, we might expect the waves to be spread sideways like a ship-wave pattern of a single mountain, but this does not appear to happen. The waves get into ladder-like arrangements, each with its own phase. The mechanisms which cause this are concerned with the separation of the flow from the ground, and with the formation of rotors, which are eddies in which the air rotates, usually under a wave crest or on the lee side of a mountain. These have the effect of altering the shape of the ground by adding the eddies to the ground shape, and with the streamline which encloses them being the new shape. Since they are formed under the influence of the

waves they tend to perpetuate the pattern initially set up by the first mountain the air encounters.

Nature works against the wave amplitude becoming too large. The first lee wave behind a mountain may be bigger than the mountain causing it, especially if it is the downslope off a plateau, but the formation of a rotor under the wave decreases the size of the second wave. Trains of lee waves off plateaus are not always visible in cloud patterns because the descent from the plateau has evaporated them. They are there nevertheless and may be detectable by other means. They can be seen in the sun glint which varies in intensity, when the sea is calmer under the wave crests and rougher under the troughs where the streamlines are compressed towards the surface.

FÖHN WINDS, FÖHN CIRRUS AND MOUNTAIN WAVE CIRRUS

If a cold airmass is blocked by a mountain range too high for it to flow over, the high-level wind may flow across and down the lee side. The descent warms the air and makes it very dry, and this may create very pleasant conditions sometimes. But when the wind is strong and possibly gusty at the same time, it makes people restless, upset, with dry throats and possibly headaches. The sky undoubtedly has a psychological effect on these occasions because the waves bring to mind previous worrying experiences.

If the cold air is capped with cloud which just manages to spill over the top it is filled with wave clouds which evaporate after a few waves as the air descends the lee slope, although the waves continue further, and with them the great variations at the ground. The resulting uncertain wind behaviour creates mental unrest. These winds are called föhn winds in the European Alps, and we use the same name for such winds in other places too.

It has been widely taught that the warmth and dryness of the wind is the result of rain having fallen on the mountains so that it is much less cloudy on the lee side and warmer because the latent heat of condensation of the rain has remained in the air. This is a good explanation because it explains the two most obvious characteristics of the air on the lee side. But it is not usually the correct explanation because the rain on the mountains does not usually occur! There are other winds with similar characteristics which come from a dry cloudless plateau, of which the berg winds of Natal are a good example. When the cloud top does not spill over the Alps and the peaks are in sunshine, rain obviously plays no part: the cold air behaves as a plateau. A south föhn blows down the Swiss and Austrian side and a north föhn down the Italian side of the Alps.

A good example of the rainfall effect is the great contrast across the mountains of Japan when a northwesterly blows across the sea of Japan with heavy snow showers on the wind-facing slopes and glorious sunshine on the plains of Tokyo. The chinook of the northwestern Rockies has many similarities, but its other features entitle it to a separate name.

It might be expected that such a ground-level effect

147

▲ 9.1 (14/2/62, 14:20 PM, SE)
Lee waves seen from above near Corfu.

▲ 9.2 (5/7/67, 21:15 PM, NW)
 A solitary wave cloud with billows on its top near Boise, Idaho.

 ▲ 9.3 Wave clouds in Greenland over mountains protruding though the show plateau.

▲ 9.4 (3/7/63, 10:30 AM, NW)
Moist air coming up the lee side of the Pic du Midi (2,885m, 9,465ft) in southern France.

▲ 9.5 Billows in a wave cloud in California.

▲ 9.6 (6/2/62, 18:00 PM, N)
The gigantic wake of Kanchenjunga (8,600m, 28,217ft at a distance of 72km (45 miles) filled with turbulent cloud.

▲ 9.7

9.7 Mother of pearl cloud seen near Oslo in a northwesterly wind, viewed looking west after sunset.

It is a standing wave cloud composed mainly of spherical droplets which cause iridescent colouring depending on the angle of the sun and the droplet size. Some droplets are freezing and emerging from the cloud as tenuous cirrus from the upper left side. This cloud presents a serious problem because at its height of 25km (15.5 miles) the temperature was -80 deg C. Normally droplets would freeze immediately; but the colours and the sharp lee edge of the cloud indicate spherical droplets evaporating there. Either the droplets are too small to freeze or are prevented from freezing by acid (probably sulphuric) which is often found in thin layers in the stratosphere.

9.8 The Owens Valley, seen here, lies next to the Sierra Nevada of California. Mt Whitney (4,418m, 14,495ft) is seen on the right. The air descends to the valley floor (at about 1,300m, 4,265ft) and then rises up into the first lee wave to nearly 5,500m (18,045ft) before forming the lee wave cloud which is partly over the Inyo Mts. The great size of the lee wave induces a rotor in which the desert dust blown up from the valley floor accumulates. This is the dustiest occasion yet recorded, and the wave was so powerful that Bob Symons, the pilot, was able to soar a Lightning twin-prop fighter aircraft with zero engine power, like a glider.

▼ 9.8

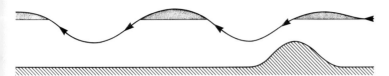

9.9 Mountain wave with lee waves downstream. Lee waves are often set up when an air stream crosses a hill and oscillates about its equilibrium level. Clouds may occur in the wave crests.

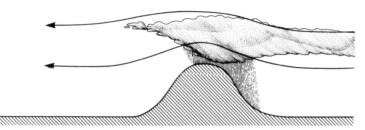

9.10 Föhn due to rain with lee clearance. The classical föhn consists of rain falling out of cloud, formed by ascent over a mountain, with less cloud and a higher cloud base on the lee side.

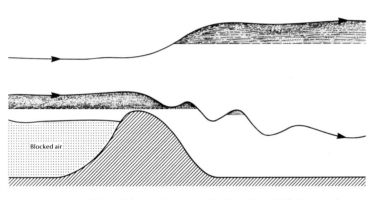

Blocked air

9.11 Föhn without rain caused by blocking. This is a much more common föhn, caused by the blocking of a cold air mass by a mountain, so that air from a higher level descends the lee slope.

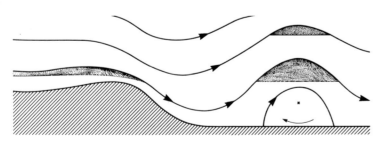

9.12 Large lee wave with rotor. When a large lee wave is generated, a rotor is formed under the wave crest, reversing the wind at the ground.

would be confined to the bottom of the atmosphere. Not so! Natural events soon find their steady-state if there is one, and a great depth of air may be involved. When part of an airstream is blocked by mountains, what goes over the top has a greater depth to fill on the other side; therefore it must be retarded to maintain the same volume flow; therefore it must travel towards higher pressure, which provides a decelerating force; therefore part of the air must be made cooler and denser to increase its weight; therefore the air in the higher levels must rise, and this produces a wave cloud over the lee slope, or quite near to it.

Thus, in the place where the low cloud is disappearing when it just spills over the mountain a high wave cloud appears and may be extended far downwind. When seen from the ground where the low cloud is pouring down the lee slope but evaporating and getting nowhere, the piece of clear sky between it and the high wave cloud forming overhead is called the föhn window (föhn lücke).

The high-level wave cloud is usually composed of supercooled water droplets at first, but it may well have time in which to freeze, and then if it is not brought down to below its ice evaporation level it may persist for several hundred kilometres as a long trail called föhn cirrus. If there is no blocking of the lower-level air, as in the case of flow over an isolated peak, a cirrus wave may still be formed, but the air would soon return to its original level. If this wave were still above the ice evaporation level a cirrus trail would be seen, but for this to happen the air would have been supersaturated for ice but not for water on the upwind side of the mountain. On some occasions, such as when a warm front is approaching, there is a layer of cirrus already, which evaporates over the mountains and reforms again on the lee side.

Perhaps the most spectacular mountain wave clouds are the long pennants of cirrus which originate in a wave over or just downwind of a fairly small mountain, but extend for many hundreds of kilometres downwind. These are the true orographic cirrus, or should we say orogenic? The more homely mountain wave or hill wave cirrus is preferred as a name by some.

Cirrus streamers are common over Iceland and many Arctic islands. The most impressive on record is one formed by the small island of Jan Mayen, which is well known for the great variety of wave and vortex patterns found in its lee. The main peak is a mere 15km (9.3 miles) across and 2,277m (7,470ft) high. On this occasion the orographic cirrus was 150km (93 miles) wide and at least 500km (311 miles) long, reaching to the coast of Greenland.

150

▲ 9.13 (13/4/86, 12:35 PM, CZ5)

9.14 (21/1/80, 4:00 PM, NE) ▶

9.13 Orographic cirrus is trailing away in the high-level jetstream from a supercooled wave cloud formed over the descent from the snow plateau of S Greenland (see the similar cloud without a trail in 7.9). The trail seems to be hanging down like combed hair; but turn the picture so that the top is on the left and it now looks like hair blown in the wind from the mountain top. Satellite pictures do not have a 'top', but are conventionally shown like a map with North at the top of the page.

The coastal ice is melting rapidly at this time of year (April) and forms variously shaped whirls, particularly in the Denmark Strait across to Iceland (on the right edge) where snow still caps the mountains.

In mid-morning we see the shadow cast by the wave cloud on the snow below at its northern edge. The cloud close to the sea surface shows traces of lee waves coming off Iceland with tiny newly formed cells in the light wind beneath the jetstream.

9.14 The Rocky Mountains at Banff, Alberta, displace the westerly winds in some odd ways: the higher levels often being displaced in the opposite to the direction of the lower layers following the ups and downs of the ground. Thus, above the descent from the mountains to the plains beyond, a large band of the middle levels rises into a lee wave, and here we see the front of that wave and the cloud is clearly seen to consist of several layers of alternately high and low humidity.

We are looking northeastwards at about 4 pm in January, and the edges of the layers are lit up by the low sun. Such plywood looking clouds are often called 'a pile of plates'. In the shearing motion of the jetstreams which are often close at hand when these clouds appear, any compact mass of moist air produced by big cumulus in the previous few days, is converted into a stretched out sheet of high humidity, and produces one of these protrusions.

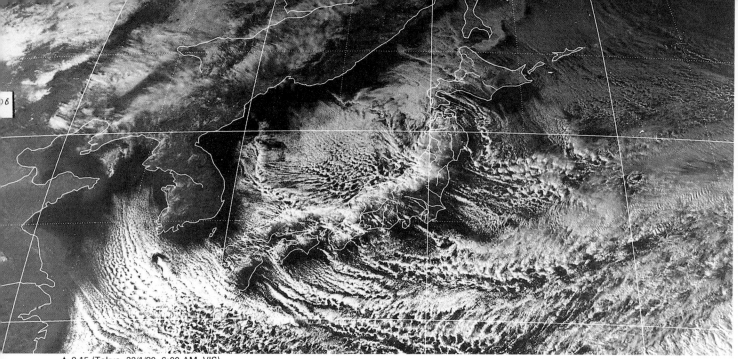

▲ 9.15 (Tokyo, 22/1/80, 6:00 AM, VIS)

9.15 The dry winter wind is blowing from Manchuria across the Sea of Japan, where it picks up moisture and deposits snow showers on the northwest side of the mountains of Japan. The deposition causes a föhn clearance on the lee side and the Plain of Tokyo is bathed in sunshine. Out over the warmer ocean the cloud is again renewed, most readily in the lines of flow that pass through mountain gaps.

In the extreme east as the air passes over the North Pacific extensive billows are formed on the cloud top under a strong inversion. Notice the clear space in the lee of Korea and the vortex street from the island Cheju-Do.

9.16a–c (a) A west wind blows across the Savoy Alps from S France into Italy. As it descends the lee slopes waves are formed. As the cloud evaporates in the descending air föhn cirrus takes over at high altitude. The north end of the island of Corsica is seen at the bottom of the picture, and by Channel 2 (near visible infra-red) we see bright glint on the calm sea in its lee.

(b) The deep infra-red (Channel 4) picture shows that the land of Corsica is so close in temperature to the sea that it does not stand out. The cold high föhn cirrus is prominent.

(c) In Channel 3 the whole sea is a-glint, and the cirrus (large particles) is black, while the lee waves (small particles) are bright. Elba and some nearby clouds are dark also, by contrast with the sea glint in the dancing morning sunshine. Some faint contrails are discernible above the dark cirrus cloud.

▲ 9.16a (20/4/87, 08:04 AM, 2)   ▼ 9.16b (20/4/87, 08:04 AM, 4)

▼ 9.16c (20/4/87, 08:04 AM, 3)

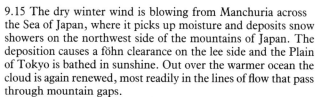

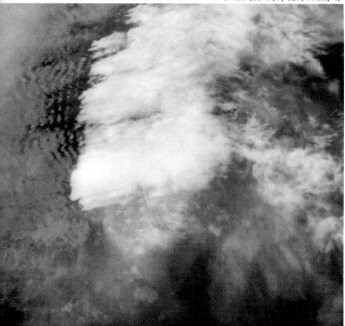

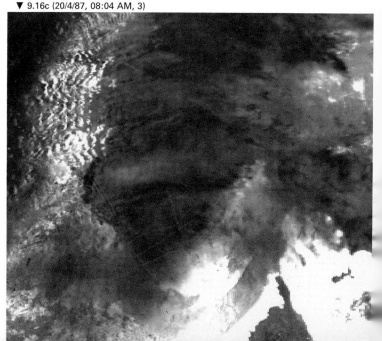

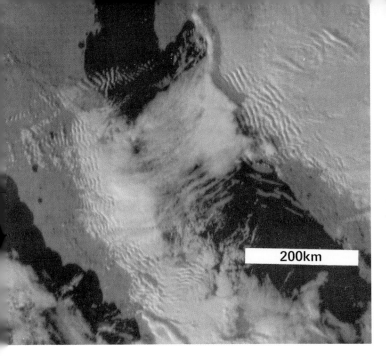

200km

▲ 9.17a (26/12/82, 13:28 PM, 2)

▲ 9.17b (26/12/82, 13:28 PM, 4)

9.17a (Channel 2), b (Channel 4). Lee waves and föhn cirrus with its shadow stretch across the north Adriatic Sea, but both evaporate on the west side of Italy. This is very typical of the Bora – the cold continental winter wind which blows harshly from the E European interior. It is particularly strong in Trieste, which lies on the bay to the north of the peninsula opposite Venice at the north end of the long coastal wave cloud.

9.18 A southwest wind blows strongly across the Western Isles of Scotland, and produces a beautiful 'ship wave' from the remote tiny island of St Kilda (425m, 1,394ft) which lies 200km (124 miles) to the west of the mainland. Further south the lighter wind produces lee waves of much shorter length which almost look like (but are not) billows over N Ireland and England, in the stratus layer which has a very strong inversion at its top.

9.18 (2/12/82, 11:37 AM, CZ5) ▶

10
ON REFLECTION

CALM

A mirror-like reflection on water promotes a sensation of peace. Perfection is elusive because the slightest gust ruffles the surface. An optical event that shines at us even though we have our backs to the sun requires some reflection, and nowhere does this occur more beautifully than on the water surface that only yesterday was whipped up into a rage.

The colours are softly exquisite; the symmetry so near-ly perfect that the small asymmetries of tone and space fascinate the perceptive eye. Perfect symmetry would be-tray itself as a moronic error, for the lower scene is not the upper scene reversed as if it were a picture on the wall seen in a mirror placed flat on the table. There is something three-dimensional which Nature adds to this one-eyed view: the baseline is not the mere eight centi-metres between our eyes, but the full five metres between the camera and its mirror image in the lough. The mind is aware of a spatial reality it cannot instantly understand. Are you sure this is not printed upside down?

▲ 10.2 (Jersey)

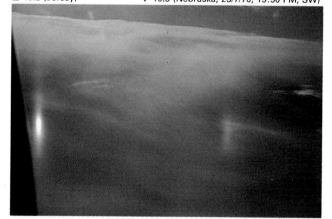

▼ 10.3 (Nebraska, 23/7/70, 19:50 PM, SW)

10.1 (*top*) (21/8/87, 9:30 AM, NE) Lough Ree (Ireland) and sky; a summer morning with cumulus growing overland.

10.2 Sun pillars are produced by reflections from the faces of hexagonal crystals with horizontal axes. Refraction through these crystals produces a deviation of at least 22 deg, and more if the axis is not perpendicular to the plane through the sun, the observer, and the crystal: this produces the tangential arc seen here at the top of the pillar. The arc is tangential to the common 22 deg halo, but this is not seen in this case because it requires the presence of crystals with all orientations, and only horizontal ones are present here.

10.3 On the left is the sub-sun, and on the right is the cor-responding sub-mock-sun, seen from an aircraft, by reflection from horizontal crystal faces and refraction in hexagonal prisms with vertical axes.

SUB-SUNS AND PILLARS

The sun reflected in a nearly calm lake produces a pillar of light. The tilting by waves throws twinkling sunbeams at our eyes from far away and from close to us on the water. But the tilt is not enough to divert a ray sideways by an equal angle when the sun is low. The pillar is widened if we stand high above the water or the sun shines more steeply down.

The fragmentation of the sun's image in the water into a diffusely illuminated area is increased by steeper waves. A wavy surface converts a narrow beam of rays from the sun into a spreading cone of light, the angle of spread widening as the waves are raised. To see glint at a particular point on the sea from a place in the sky we must be within the spreading cone of rays from that point.

Glint indicates the state of the sea. Where the sea is flat calm the satellite will report it as being either very bright with the reflection of the sun, or very dark because the satellite was not in the reflected beam from that bit of water. If the sea at our point of interest is very rough the cone of scattered sunlight will be very wide. If the sea nearby is less rough it will appear brighter than the rough area if the satellite passes through its cone of scattered sunshine, or darker if the satellite does not.

Glint is a sort of sub-sun, which is the name given to the refelection of the sun in the flat tops of ice crystals whose tops are horizontal. Slight deviations from the horizontal cause a spreading of the image, and as with the pillar there is much less spread sideways. We do not see a pillar when the sun is high.

10.4 The camera was tilted so that this picture could include both mock suns on the parhelic circle and the upper half of the 22 deg halo.

10.5 This shows a phenomenon rarely seen even from aircraft – the white horizontal circle through the observer's shadow (the anti-solar point). It is equivalent to a reflection of part of the parhelic circle in horizontal crystal faces and it was difficult to photograph because of the steep downward angle of view.

A pillar is seen in the sky when the sun is low. A mock sun (described in Chapter 11) lies on the parhelic circle, which is a circle of white light seen around the sky at the altitude of the sun when there is an abundance of ice crystals with vertical faces. This illustrates how, if the crystal faces are restricted to one plane, we get a spot of light, but if they can take up any direction but with the crystal axis direction fixed, we get an arc which is perpendicular to their axes. It is rare for the parhelic circle to be complete or very bright. Just as the sub-sun can produce a mock sub-sun, so a sub-sun can produce a sub-parhelic circle. The reflection does not have to be in the same crystals as those that make the sub-sun, although it could be.

On rare occasions the sub-sun appears to be surrounded by a ring of light, called Bottlinger's ring. The eye increases the contrast between the duller area surrounding the central spot and the outside of the ring. It also intensifies the contrast between the ring and the dull area outside it. Occasionally it creates an illusory dark region in the middle of the central spot.

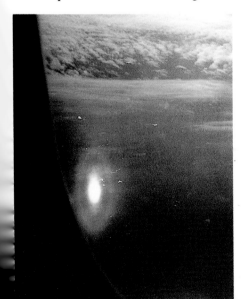

10.6 Bottlinger's ring requires the crystals reflecting the sub-sun upwards to be oscillating in such a way that, like a pendulum, they spend more time at the ends of the swing, to generate the outer ring, and more frequently pass through the central (horizontal) position because the swinging in all directions contributes to the same centre. The absence of a second discrete ring means that the theory of a downward second reflection followed by a third, upward one is not correct in this case.
10.7 To see a sun pillar beneath the sun one has to be above and close to a cloud of ice crystals.

▲ 10.8 To see the reflection of a light source in the horizontal bases of ice crystals requires a very strong light source at night. This was given by a Japanese fishing fleet off the north coast of Japan and observed from the city of Fukui in Honshu. A more extensive display had been reported in 1983.

The sub-sun is produced by light reflected from plate-like crystals with horizontal surfaces. If these plates lie very close to the horizontal in the air they produce a small spot of light elongated towards and away from the sun. But if they oscillate around the horizontal position the bright central spot remains because it is produced by oscillations in all directions. As the oscillations increase a ring is produced by crystals at the limit of their oscillations, which is brighter than the reflections intermediate between

the ring and the central spot because the crystals spend less time in the intermediate positions – like a pendulum which spends more time at the limit of its swing than in the middle of the swing where it is moving rapidly.

The sub-suns and the ring are best seen in very tenuous cloud because the brightness surrounding the ring is much less than in dense cloud.

It has been suggested that rings may be due to a second reflection of the rays of the sub-sun on the bases of the crystals and then back up again by a third reflection on the tops. Such rings are theoretically possible but that explanation is definitely not correct for the case we show.

INTENSE SCATTER

A big collection of surprises came from Channel 3 of the satellite radiometer. The channel was chosen because it has a wavelength which is not absorbed by water vapour, and therefore it would be good at seeing down to the surface, and has a long enough wavelength to see through haze.

Relative to scatter from clouds and other objects Channel 3 is very strongly reflected at a water surface and is very strongly absorbed on passing through water. Ice is thought to have very similar properties to those of water for Channel 3, but we have no evidence that it is reflected to the same extent at the surface of snow or ice crystals because most ice clouds look intensely dark, as do snow-covered surfaces and frozen sea.

Even the very near infra-red wavelengths can see haze, but Channel 3 sees through it. An interesting consequence of this is that Channel 3 makes very good shadows. These are so sharp that the shadows of cirrus (which are very dark) are like duplicates of the clouds. Invisible light shadows are softened by the diffuse illumination of skylight which comes from clouds and haze. There is virtually no skylight in Channel 3.

10.9a (Channel 2), b (Channel 3) These show glint on the sea around the straits of Gibraltar. By Channel 2 we see bright patches at Malaga and on the opposite Moroccan coast where the sea is calm. The light east wind emerges from the straits as a strong jet which darkens the glint by spreading it through a wider angle. The calm area around Faro on the south coast of Portugal reflect a very narrow beam which was not intercepted by the satellite, which therefore shows it as blacker than the sea over a very wide area. The Channel 3 picture shows the same features much exaggerated, and the contrast is very great between the Faro area and that to the east of it. Lee waves can be seen in the lee of the Cape Tres Forcas at Melilla on the Moroccan coast, the wavelength being about 5km (3 miles). In the areas of glint the clouds look dark.

▼ 10.9a (28/6/82, 14:25 PM, VIS)

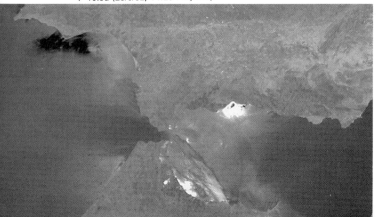

▼ 10.9b (28/6/82, 14:25 PM, 3)

11
LIGHT WORK

RAINBOWS AND CLOUD BOWS

We now come to the phenomenon of refraction and the production of coloured arcs, together with diffraction, which is the distortion of the light wave fronts by passage across obstacles whose size is in the same range as the wavelength of the light.

Newton's famous investigation of the splitting of light into the colours of the rainbow by passage through a glass prism, and the reconstitution of white light by combining them again is a wonderful example of probing the mechanisms of nature by simple experiments.

In the morning sunshine on wet grass we can observe individual drops of water transmitting brightly coloured rays to our eyes, and as we move the colour changes from blue to green, orange and then a gorgeous purple. This whole sequence can be captured in one picture by taking a close-up photograph of the drop with the light beam in the centre but put out of focus. This spreads the beam out across the film to display the colours seen from different directions. In the picture we can see several supernumerary bows which appear inside the main (primary) bow when the drops are all close to one size. This condition is satisfied perfectly in the picture because we have only one drop size!

The secondary bow, in which the colours are reversed, is seen in the neighbouring drop, with the darker area in between. If you can repeat this experiment in your back garden one sunny morning you will be able to see why the colour bands of the secondary and primary bows are not parallel.

It is interesting that the droplets on the grass are of two kinds. Along the blades are many very small droplets which have been condensed from the air when the grass was cooled by radiation into the night sky to a temperature below the dew point of the air. At the tips of the blades are single large drops produced by guttation, which is the exudation of water from the grass to keep it cool in the sunshine by evaporation. During the night the blades become very cool and so there is no purpose served by this, and the water does not evaporate and accumulates as a big drop. This is caused by the roots remaining warm in the soil (which became very warm the previous day in sunshine) and taking up water from the soil, exerting an internal pressure to exude it at the top.

We normally see only part of the upper half of a rainbow, and even at sunrise or sunset we see only a semicircle at most. At the centre of the circular arc is the shadow of the observer's head. There are two ways in which we might see more from ground level: by observing the bow formed by the reflection of the sunshine in the water surface, which is equivalent to the sun being below the ground and our shadow up in the air, or by putting ourselves above the rain. The former is a rare opportunity because most good rainbow days are showery with gusts of wind which ruffle the water and cause the reflected sunshine to be scattered, and we need a concentrated source to get a bow.

We can, however, get above the droplets in an aircraft, and if we choose an old piece of altocumulus in which the smaller drops have evaporated and the largest have fallen out (ie a thin cloud with very little vertical motion in it) we may see a coloured cloud bow. Most cloud bows are white because of the large range of drop sizes, but they can be made more intense by viewing through a polarising filter, especially when they are faint, as in the top of stratocumulus over a warm surface (because of the great variation in drop size).

Often when flying above cloud we see a glory, which consists of coloured rings around the shadow of the aircraft, and so it is likely to be seen at the same time as the cloud bow. It is caused by diffraction. The radii of the coloured rings depend on the drop size and the wavelength. With one drop size the colours may be spread out in three or four concentric rings, but with drops of a narrow range of sizes and on close observation by eye the rings are seen to be spread out almost as far as the innermost supernumerary cloud bow. Bows seen in fog are most often white and are best seen when the fog is almost transparent with a dark background. This is virtually impossible to capture on film, but it is an exhilarating experience to see!

Diffraction is much more intense in the forward direction of the light beam, but then it is important to obscure the sun to get a good view of the corona, as it is called. Whisps of thin cloud carried across the top of a mountain produce very good coronas. The same phenomenon can be seen in clouds even 20 degrees or more from the sun or from the observer's shadow. It is then called iridescence. The best examples are in wave clouds where there is no turbulence to mix different bits of cloud or cause condensation by relative upward motion. The drop size grows as the wave crest is approached and beyond that it decreases again, but keeping the drops in any bit of cloud much the same as each other. The colours are then much purer and vary with drop size and angular distance from the sun (or the observer's shadow).

11.1 Cloud bows are best seen on thin cloud that is dark because it has lost its smaller drops with time, and does not scatter bright white light.

Bows are not usually as obvious as this one and the fact that they are circular, surrounding the aircraft shadow (and a glory) is made difficult to recognise by the fact that the eye is looking at something laid out on a horizontal surface – the cloud top – so that in 3-D it looks like a hyperbola which is the section of the cone from the eye to the arc by the plane of the cloud top. It therefore does not 'look' circular.

Its detection is greatly helped by the use of a polarising filter, because the light from the arc is highly polarised.

11.2 A fog bow is, of course, a cloud bow. But the fog has to be well illuminated from behind the observer, as is this one which is viewed on shallow fog from the top deck of the QEII.

11.3 Bows, arcs, etc on wet grass are affected by the forms of water drops present. Guttation is the unevaporated perspiration at the tip of the blade forced out by the pressure generated in the roots which are at a higher temperature late in the night when the surface is cooled down to the dew point. That large drop compares with the multitude of small ones, which are dew condensed from the air on to the cooled blade of grass.

11.4 To capture the rainbow spectrum emitted by individual drops we defocus a closeup view. The drops on the right show

the main bow with the supernumeraries inside; those on the left show the secondary rainbow outside the main arc with the order of the colours reversed. The lack of co-ordination of the drops depending on size, distortion from spherical and distance from the eye means that only very smudged bows are seen on wet grass, or even on wet spiders' webs on grass.

11.5 The glory surrounding the observer's shadow is best when only a small range of droplet sizes is present. In a cloud with a very narrow range of droplet sizes, coloured arcs fill the whole space between the cloud bow and the glory, but only a very alert observer will see their fleeting appearances.

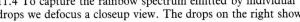

11.6 The circumscribing halo is almost elliptical when the sun is high, as here. It is produced by hexagonal prisms with horizontal edges. The common halo can be seen inside it, particularly in the 3 to 6 o'clock sector: this is generated by prisms with edges randomly oriented in three dimensions.

The Lowitz arc which is tangent to the 22 deg halo is visible at 4 o'clock: it is very rare and is generated by prisms whose sections are irregular hexagons, which are very rare.

The right side of the contrail which crosses from 140 deg to 320 deg is sharper than the other edge because its shadow darkens the cloud there.

11.7 Rare events seem rarer than they actually are because they so often pass unnoticed. This picture contains a 10 deg and a 17 deg halo on the outside of an 8 deg halo. These are all produced by the passage of rays from the sun through the pyramidal ends (which may be protruding or hollow) of hexagonal prisms. Such prisms usually have flat ends perpendicular to their axes and so these are indeed rare. Since there are many different paths possible through two faces of such a crystal many different

(cont'd p162)

▲ 11.6 (Brookhaven, Long Island)

▲ 11.7 (Wimbledon, 14/4/74, 19:30 PM)

▼ 11.8 (Wimbledon, 5/3/61, 16:45 PM)

11.9 (Mt Olympus, Washington, 16/1/66) ▼

HALOS – REFRACTION IN ICE CRYSTALS

Ice crystals can take many different shapes, the most common of which are illustrated in the diagrams showing the paths of the light rays which produce the various arcs which may be seen.

The most common phenomenon is probably the sun dog or mock sun, technically called a parhelion. When light rays pass through two crystal faces at 60 deg to one another they are deviated by at least 22 deg, this being achieved when the symmetry of the path is complete. This means that such rays cannot reach the eye from a direction less than 22 deg from the sun. Therefore, in a cloud composed mainly of hexagonal prisms of ice which is thin enough for rays to penetrate with only one encounter with a crystal, the sky is darker within this angle with brighter cloud outside from which light can be deviated to the eye. Because red is deviated through a smaller angle than other colours the dark area is smaller and the edge is tinged with red. At slightly greater angles the colours are superimposed because most of the crystals deviate the rays somewhat more than the minimum. Outside the red ring the sky is a brighter white, and this constitutes the common 22 deg halo. To form a complete circle there must be crystals in the cloud oriented at all angles.

But some crystals fall with their axes preferentially vertical or horizontal. Those with vertical axes produce bright spots close to the halo with red and orange often very bright on the side towards the sun and an extension on the other side which is white. These are the mock suns, and when the sun is low they are almost on the

22 deg halo, but with higher solar altitudes they are well outside the halo. The crystals responsible are probably kept in the vertical position by a flat plate on the top acting as a parachute.

Some of these vertical crystals reflect, or refract, the sun's rays without altering the inclination to the horizontal and this produces a white circle at the same altitude as the sun, called the parhelic circle.

Another group of crystals having their axes horizontal produce the top point of the halo when their axes are perpendicular to the ray, but otherwise produce points on the tangent arc, either at the top or bottom, the upper and lower tangent arcs having shapes which are very dependent on the altitude of the sun. At high altitudes around 50 deg above the horizon the tangent arcs are joined together to form the circumscribing halo, which is approximately elliptical. At low altitudes the arcs are concave away from the halo at the point of contact and do not join up. The upper arc is often seen together with a sun pillar when the sun is low. The lower arc is best seen from the air looking down on the cloud, and forms a loop at low sun altitudes. At altitudes around 22 deg it becomes mixed up with the sub-sun.

Hexagonal prisms of ice often have 90 deg corners at the ends. This means that there may exist a whole system of halos for these corresponding to those generated by the 60 deg prisms already described, although they are usually much less bright and less common because the areas of the relevant faces are less. They produce the 46 deg halo and its associated arcs.

More interesting, perhaps because they are more rare, are the halos generated by hexagonal crystals with pyramidal ends. Although this crystal form has been known for a long time its appearance seems much rarer than the others. There is a hint in the circumstances of occurrence of the 8 deg halo that they are formed at higher temperatures than the others and are near the ground on most occasions. The pyramidal ends may be hollow, but that would make them produce less bright results. The 8 deg halo uses a pyramidal face and the opposite prism face, the 17 deg halo uses two opposite prism faces at one end, and these are likely to be the brightest. The 19 deg halo is formed by rays which pass in at a pyramid face at one end and emerge from the pyramid face at the other end which is neither adjacent nor opposite on the hexagon.

Many other halos are theoretically possible, but many reports are not accurate enough for any explanation to be certain. In any case some of the rarer halos have not been reported often enough for there to be a proper definition of precisely what measurement to make on a halo whose limits appear rather broad.

Only those with a special interest in the trigonometry will find that part of the subject of interest before they have seen one of the rare halos themselves. Very few have been reported as being brightly coloured, and so now we turn our attention to that possibility.

BRIGHTLY COLOURED ARCS

The rainbow is the most brightly coloured of all the optical phenomena because it is well illuminated by the sun and

---

(captions from p161 cont'd)

arcs are possible. The outer ones are likely to be taken for the common 22 deg halo by less well-informed observers, but the 8 deg halo stands apart from the others, and has been more often reported. The reports do not contradict the theory that the pyramidal crystals occur when the temperature is not far below 0 deg C. Freezing in clouds usually occurs in the region between -20 deg and -40 deg.

11.8 The circumzenithal arc is a truly circular arc subtending an angle up to about 90 deg at the zenith, its centre. The colours are very clear as in a rainbow, against a blue sky. The prisms are 90 deg between the vertical sides and top ends of the crystals, and the arc can only be seen when the sun is less than 32 deg above the horizon. At a solar altitude of 22 deg the arc is tangent to the 46 deg halo (see 11.10) but it remains within 1 deg of that halo at its nearest point for a solar altitude between 16 deg and 27 deg.

In this picture it happens to be tangent to a contrail, although the crystals forming it could be at a different height.

11.9 When the sun's rays are refracted through the lower 90 deg edges of prisms with vertical sides an arc parallel to the horizon, with very bright colours is produced. This requires the sun to be within 32 deg of the zenith; it therefore cannot be seen in latitudes greater than $54\frac{1}{2}$ deg. The optics are the same as for the circumzenithal arc (which uses the other ends of the same kind of crystals); thus it is best when the sun is between 16 deg and 27 deg from the zenith, and would be tangent to the 46 deg halo if it were present.

This circumhorizontal arc was observed close to Mt Olympus, near Seattle.

162

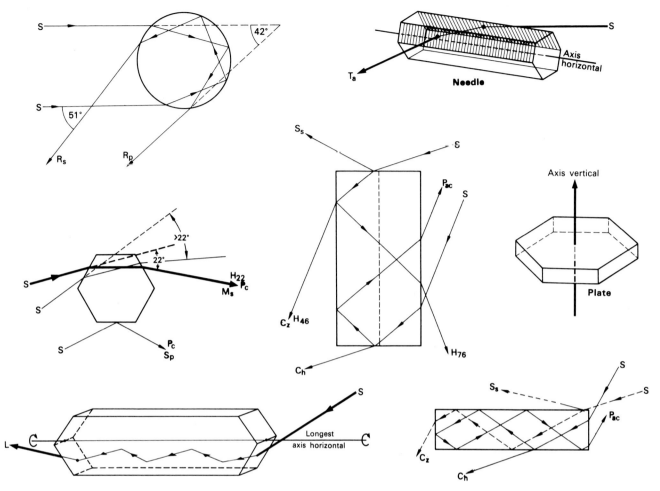

Ray paths for optical phenomena

often has a dark background. But it is not the purest of arcs: that honour is reserved for the circumzenithal and circumhorizontal arcs. In these the green is very pure, and so is the blue except for the reduction in its prominence by the background (and foreground) of blue sky often seen at the same time. Arcs are formed in the 90 deg corners of hexagonal prisms whose axes are vertical and end-faces horizontal. These prisms deviate the sun's rays through a definite angle to the eye of the observer and so the colours are pure. The arcs are circular with the zenith at the centre. For the circumzenithal arc the rays enter the crystal at the top and emerge from a side. This is possible only if the sun is below 32 deg from the horizon, and is best when it is at around 20 deg. For a large part of the range the arc is close to 46 deg altitude, and therefore looks like a tangent arc of the 46 deg halo; but as it is produced by a more restricted group of crystals its colours are purer. It subtends at most about 90 deg at the zenith (its centre) and has red towards the sun.

The circumhorizontal arc is formed by rays entering at the side face of the hexagonal prism and emerging from the horizontal base. It can only be formed when the sun is within 32 deg of the zenith, and therefore can never be seen polewards of latitude 54 deg. It is best with a solar altitude of about 70 deg. It is best seen from high mountains because it requires the crystals to be lower than 32 deg above the horizon.

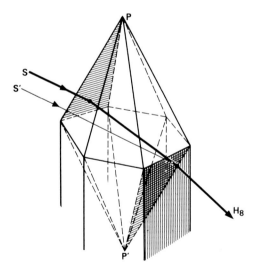

C_h	=	Circumhorizontal arc	Rp	=	Primary rainbow
C_z	=	Circumzenithal arc	Rs	=	Secondary rainbow
H_o	=	0° Halo (random axes)	S	=	Sun
			Sp	=	Sun pillar (horizontal axes)
L	=	Lowitz arc			
Ms	=	Mock-sun (vertical axes)	Ss	=	Sub-sun (horizontal faces)
Pac	=	Anti parhelic circle	Ta	=	Tangent arc (random horizontal axes)
Pc	=	Parhelic circle .			

163

All halos, arcs and spots of coloured and white light seen through ice crystals in the sky are fascinating details and tell us a great deal about the tiny particles which cause them to be seen. The enthusiastic observer must be warned that all sorts of optical effects are produced in camera lenses exposed to direct sunlight, and it is desirable to make notes and sketches of what was seen rather than rely on photographs to tell the story. After all the picture may not turn out to be satisfactory.

▲ 11.10 (South Pole, 13/1/86, 01:47 AM)

11.10 The 46 deg halo is formed in the randomly orientated 90 deg edges of crystals. It is not common and the best pictures have been obtained in Antarctica, like this one at the South Pole. The upper tangent arcs of both the 22 deg and 46 deg halos are present. These are produced by refraction through crystals with horizontal edges, but with the faces not vertical or horizontal except at the point of contact with the halo. The halos and the arcs are red on the side towards the sun but the other colours are smudged by superposition, as with the mock sun.

The 46 deg halo is rare because most usually the crystals with the 90 deg edges have vertical axes, which means that the circumzenithal arc may be present, and at its points nearest to the sun is superimposed on the tangent arc. Cases in which the tangent and circumzenithal arcs can both be separately identified are very rare indeed.

▼ 11.11

11.11 The mechanism which separates the positive and negative electric charges in thunderclouds has long been discussed. The agents of the separation are certainly the falling particles, of which the hail is certainly the most effective because they most efficiently retain small supercooled droplets with which they collide. The electric field always exists to some extent and this ensures that the large faster falling particles capture more particles of one sign, but the description of the best informed opinion about the details of the mechanism is outside the scope of this book.

The dramatic nature of lightning discharges continues to thrill, as it has done since animal life began, because of the simultaneous noise, heavy fallout, gusty winds, and bright flashes.

This picture shows the deep illumination of cloud and fallout by cloud to cloud flashes after sunset. In interpreting the tortuous tracks of the discharges it must be remembered that they are extended in three dimensions, so that the curvature is exaggerated in the flat picture.

12
FOGS AND TRAILS

FOG IS BEAUTIFUL

In previous chapters the clouds described have all been formed as a result of the cooling of air by ascent through the atmosphere. But there are other ways of forming cloud. Fogs are formed at night when the sky is clear and there is no blanket of cloud to keep the Earth warm. There is also the possibility of clouds being formed by the mixing of two masses of air at different temperatures, but this usually requires rather extreme temperatures. And finally, there is the possibility that by some quirk there are not enough condensation nuclei to form a respectable cloud. What then happens when the nuclei are supplied?

We need no false analogies about white blankets cosily clothing the countryside nor talk of the sun's glare being softened so that we can stare it in the face. Fog IS beautiful. It is the calm antithesis of storms. They climb high in the sky to manufacture their hail, twisters and electric threats; fronts spin up cyclones to wreck ships at sea, blow down forests, flood habitations, and erode soil. Fog rests without shame, humbly upon the Earth

In the peace of a foggy morning that elusive substance – cloud – is there for us to walk into and breathe into our being. It holds no secrets from us but holds us secret from the world. The wet air, like dew on whiskers, clothes the September spiders' webs in beads which shine with spectral colours when the sun breaks through. It brings us close to things that are near without the distractions

12.1 Calm air can only be cooled significantly by radiation because the conductivity of air is negligible. Heat has been lost downwards by the air while most of the energy lost upwards by the ground is in wavelengths not absorbed by air and therefore escapes into space. The ground becomes cooler than the air when the sun is low or below the horizon and so the air close to it is cooled, to produce mist. Calm clear autumn mornings and evenings are the best time of year for this because of the evapotranspiration of moisture by the lush vegetation.

of the inaccessible distance.

Fog quietens the distance and softens the noises just beyond vision, making them curious with uncertainty; and we explore what is near with fresh power of conscious discovery and less haste to leave it behind.

Fog stays at home and in doing so is made to accompany the stagnation of domestic effluents and industrial bad odours carelessly thrown out by thoughtless planning. Yet it assumes a cleansing role, capturing each small particle in a droplet which will carry it to the ground, dissolving acid molecules likewise. Men meanwhile blame it for the accumulation of their effluents. When it vanishes like a ghost before the sunshine it unpacks for us the present of another always different day which beckons us with charm to smile away the delay, the sunshine coloured with warmth.

February's frost gathers the fog in geometric designs onto windows, and everywhere the bushes are clothed

165

in white. We may proceed from the holistic view, to each twig, and down to microscopic detail, each scale reminding us of the subtle motions of the air with which Nature placed these crystal jewels on each with individual detail, generating tantalising near symmetry which makes no snowflake the same as any other. We call the differences chance, while at the same time, to hide the limits of our understanding from ourselves, we assume the samenesses are caused!

Those plans we made to hurry through this day are frustrated, and because anger is clearly inappropriate we learn that to be always hurried is to miss those moments when Nature calls a halt to the tyranny of plans, especially those which bring no surprises. Fog gives us time when we thought we had none to spare.

RADIATION FOG

Every object radiates energy all the time. The amount is determined by the temperature and we can tell whether an object is hotter or colder than the surface of our hands simply by holding it near them. Our hands get warmer or colder according to which way most energy is flowing. The radiation of energy does not stop when two bodies of the same temperature are close to each other, although it appears to because nothing seems to change.

The radiation is in a restricted band of wavelengths, and very hot objects radiate short wavelengths like ultra-violet and visible light; cooler objects emit only long wavelengths which we call infra-red. Objects much colder than the Earth radiate in radio wavelengths.

Gases have a special property: they absorb and emit only in certain very restricted wavebands. They are completely transparent to all other wavelengths. When the radiation passing through the air (eg light from the sun) is analysed by noting what wavelengths are absorbed, we can tell what gases are present. The gases with the greatest absorbing power are those with more than two atoms in each molecule. Nitrogen and oxygen, which make up over 98% of the atmosphere, absorb very little of the light from the sun or of the infra-red from the Earth. Argon, which is 1% of the atmosphere, also absorbs quite negligibly.

The most absorbent gases are water vapour (H_2O), which is much more plentiful in the lowest layers of the atmosphere, and carbon dioxide (CO_2), which is very well mixed throughout the air up to at least 30km (18.6 miles). In fog, water vapour is by far the dominant atmospheric constituent.

When the sun is low or below the horizon the heat by radiation from the Earth in the wavelengths which pass through the air makes it colder than the air by anything from about 2 deg to 20 deg C. Sometimes we have ground frost, with dew freezing solid on blades of grass and car roofs while the air is much warmer. If the air is very moist the lowest layers lose heat by radiation to the colder ground, while the layers above are not in direct exchange with the ground because the intervening layers absorb all those wavelengths, and the gradual evening out of temperature by radiative exchange of each layer with its neighbours above and below is very slow. When the bottom layer which does exchange with the ground is cooled below its dew point a thin layer of ground fog or mist in patches is formed. We can see how deep the layer which has this direct exchange with cold ground is because it is about the same as the depth of the fog.

Even a slight wind causes turbulence enough to mix the cooled layer into a much deeper layer. With stronger mixing the cooling may be spread into a deep enough layer for none of it to be cooled enough to form fog. With still more mixing the result may be the formation of a layer of cloud at the top of the layer, which is the coldest part of the layer when it is well stirred and the lapse rate made equal to the dry adiabatic. Such cloud is called turbulence cloud.

The formation of fog is usually prevented by the existence of a layer of cloud, which radiates to the ground almost as much as the ground loses by its own radiation, because it is at almost the same temperature. The layer of cloud does radiate upwards through the clear sky above, and the cooling of the cloud top is communicated down by cold convection to the air and ground below, but because a much greater mass is being cooled the drop in temperature in one night is much less than if the cooling affects only the top surface of the ground and the thin bottom layer of air.

In the cold weather of winter the amount of water vapour that can be held in the air is much less, and the nights are longer, so the cooling of the ground under a clear sky is much more severe than in warmer months. The depth of air having direct radiative exchange with the ground is also much larger and so the phenomenon of very shallow layers of ground fog is much less common than in autumn, or even summer.

STEAMING FOG

When cold air moves over unusually warm water it appears to 'steam'. The same thing happens in a cold bathroom when the bath is filled with hot water. The water is so warm that as it warms the air in contact with it it can evaporate into it. Thus, if air at 0 deg C moves over water at 10 deg C this thin layer is quickly warmed to about 10 deg C and then saturated with vapour so that it contains about 7.6g/kg of air. The air just above it, still at 0 deg C would contain about 3.8g/kg if it were almost saturated. When the warm air rises by its buoyancy into the cold air and mixes in equal amounts we have 11.4g of vapour in 2kg of air at 5 deg C. But at 5 deg C air can only contain 5.5g/kg, so in the 2kg of air 0.4g of water must condense as steaming fog. It is about as much water in droplets as would be found about 70m above the condensation level in a cumulus cloud, which is quite dense cloud. We would get about the same amount of water as droplets when the cold and warm air were mixed in the ratio of 1:3 or 3:1, but less with smaller or larger proportions. Since the air higher up is usually not saturated, as the mixing proceeds into higher layers the cloud evaporates. This is essentially a shallow phenomenon with rather patchy cloud density.

In the bathroom the water is much hotter and so the cloud can be much denser. Such fogs are seen over warm rivers, such as those which have been warmed by power station waste heat, and over lakes at the end of a warm summer when cold air drains over them at night from cooled hillsides. Near the coast of Norway where the Gulf Stream reaches its most northerly point, steaming fog is seen when very cold Arctic air moves over it. The name Arctic smoke has been given to the fog.

CONTRAILS

The exhaust of an aircraft engine is very hot and contains a high proportion of water vapour (from the combustion of the hydrogen in the fuel) by atmospheric standards. At the distance of about two wing spans behind the engines the exhaust has been diluted by about 100 times its volume of surrounding air. The details are displayed in the diagram in which the coordinates are the temperature and the humidity mixing ratio. All that takes place as the mixing dilutes the exhaust is shown by the way the point representing the mixture moves in the diagram, on which is drawn the saturation curve for a particular air pressure, in this case 250mb.

All we have to note is that the exhaust starts at the point E which is about three page widths off the right of the diagram. As it is mixed with the air it moves along the line joining E to the point, representing the air. If, on the way, it crosses over the saturation line, it means

that cloud is condensed because there is too much water there for the air to contain as vapour. From this it is evident that if the air point is at A_1, below the tangent from E to the curve, no cloud is formed, but that if the air is represented by a point in the shaded region at A a trail of cloud is formed when the dilution gets to C and that it will disappear again when the dilution has proceeded beyond D.

Some trails obviously last for many hours, and so we need to look at the last part of the dilution process in more detail. The second diagram is therefore an enlarged version of the lower left corner of the first. The ice saturation line now becomes important, for if the droplets that have condensed in the trail are frozen they will not evaporate if the point remains in the doubly shaded area. If the air point is at A, a frozen trail will not evaporate.

The shapes of contrails are often quite fascinating because of the motion in the aircraft wake. The aircraft stays up by pushing air down, and the exhaust gases go directly into this downwash. When the plane has four engines the exhaust from the two outer engines is drawn into the vortices which trail back from the wingtips. They often then appear as tubes of clouds which persist when all other parts of the trail have evaporated. This is because the centres of the vortices are the last part of the aircraft wake to be mixed with surrounding air. Also, because of

(continued p170)

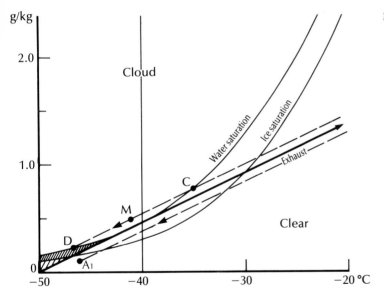

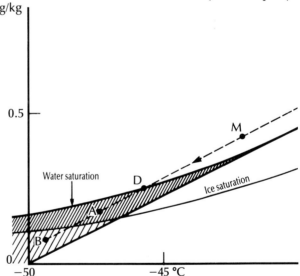

12.2 We have drawn this diagram for the 250mb level around which persistent contrails are most likely to be formed. At lower levels it is generally too warm and at higher levels it is usually too dry.

The curves show the water vapour content when the air is saturated at temperatures where contrails are likely. Clear air is represented by a point below the curves, and cloud above. In between, a cloud made of ice crystals will persist, but one made of water droplets will evaporate. Ice clouds will not form in clear air but have to be made by forming water droplets (at a point above the water curve) which then freeze.

Aircraft engine exhaust is very hot and contains a lot of water vapour. It is therefore represented by a point far off the diagram in the direction indicated (we would have to extend the diagram about three page widths to include it). As it mixes

with the air around it the point representing the mixture moves down the line from the exhaust point towards the air point. At any point M above the water curve a trail cloud would be formed. Condensation would begin at C and the cloud would evaporate beyond D.

If the air point were below the tangent from the exhaust point to the water curve, at A_1 for example, no trail would be formed, even if the line went above the ice curve.

If the air point were in the doubly shaded region, at A, because the temperature is below -40 deg C the droplet cloud at M would freeze immediately, and so the cloud would not evaporate at A. If the air point were at B the cloud would evaporate but would take quite a long time to do so because the ice crystals might have grown quite large as the mixing proceeded in the doubly shaded area.

167

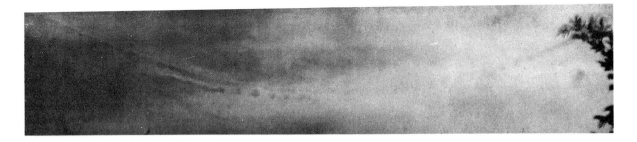

12.4 ▲ (Wimbledon, 23/9/52, 11:30 AM)

12.3 (pp 168–9) Sunset over Georgian Bay (Lake Huron) looking west-southwest, early August 1985.

In the centre of this picture is a white contrail which is higher than the cirrus, most of which either has taken on an orange-red tinge or already lies in shadow. The central band is composed of ice particles which are falling or have become much smaller by evaporation as they have fallen into rather drier air and have accumulated at that level. The wind shear has spread them into a long strip of cloud.

The blue sky in the west in the higher clear air throws the reddened cirrus into beautiful relief, and the wave clouds are darkened in the shadow of more distant clouds.

12.4 The blobs, into which the downwash curtain is changed as it is retarded in its downward motion by the stability of the surroundings, are formed whether or not there is a visible contrail. This is why a distrail sometimes appears as a row of holes 'punched' in a thin layer of cloud.

12.5 Sometimes contrails partake visibly of other motions in the air. They have been seen to take up the shape of a mountain wave. In this case the trail was formed in or very close to a layer of cirrus in which billows were developing. One part appears to be wound up in the billow vortices, but the winding did not go further than is seen in the picture.

◀ 12.5 (Beamsville, 7/12/87, 16:15 PM)

the rotation and the centrifugal force it produces, the pressure in the centre is reduced and this has the same effect as lifting the air to a lower pressure, and causes condensation. But after a minute or so, when the trail is between 13 and 26km (5 and 10 miles) long, the vortices develop loops and then appear to burst, after which the trail vanishes.

The downwash is a mixture of air and exhaust which is buoyant because of the high temperature of the exhaust. It is therefore subject to an upward force although it has been pushed downwards. The result is that the upper layers of this downwash are peeled off as it descends, leaving behind a curtain of cloud with its top along the track of the aircraft and its lower edge descending with the vortices. The vortices become stripped of their downwash air and the mixing then penetrates to their centres.

This process goes on in clear air if the contrail cloud evaporates at an earlier stage: indeed the trail tells us a lot about what goes on behind every aircraft in flight all the time. When the vortices begin to burst the trail becomes a row of puffs, and if there happens to be a cloud at that level the previously invisible downwash punches a row of holes in the cloud. Sometimes a distrail is produced in a cloud if the (invisible) downwash curtain penetrates through the cloud. This prompts the question – why does

the distrail not abolish the contrail? The answer to this question is connected to the freezing of the trail, and so we offer a picture of a frozen trail under a distrail. This, however, raises new questions because we know that the passage of an aircraft through a supercooled cloud can cause the droplets to freeze, as we illustrated in Frontal Clouds, p 29.

The action of the vortices can be seen in the case of a four-engine plane, because the trails of the inner engines can be seen to rotate around those of the outer engines. A three-engine plane leaves a central trail not knowing where to go!

Finally, the persistent trail curtain is subject to the shearing motion of the wind which moves the top along with a different velocity from the bottom. The once vertical curtain of cloud is therefore stretched out into an almost horizontal very thin sheet of cloud.

Occasionally the trail appears to become twisted. The cause of this has been the subject of much speculation, none of which has been very convincing!

TRANSFORMING SEA FOG

Sea fog used to be easy. It was called advection fog, which meant that it was formed in air which was moved

170

to a place where the sea temperature was below its dew point. When we find sea fog it is nearly always in a state of cellular convection, which means that the cooling from the top is the controlling influence and the sea is merely stopping the temperature falling below the sea temperature. If you fly across a coast early in the morning it is often possible to see where the coastline is by a change in the cellular pattern, which has more vigour over the sea because the land has less capacity to keep up the supply of heat at the bottom all night.

Satellite pictures cannot reveal the cellular patterns when they are smaller than the pixel size, but we can see evidence of the cooling at the top sometimes by the little vortices in the cloud top where the air is converging towards the top of the downcurrents.

Sunshine certainly penetrates the cloud to warm the land and some time after sunrise the texture of sea fog that has drifted across the coast shows that the invigorated convection has increased the proportion of smaller cloud droplets in the top of the cloud, which looks brighter as a consequence.

There are some diurnal variations in cloud over the sea which we discussed in Internal Gravity Waves, p 142, but the explanations given there invoked gravity waves and subsidence. But fog, being on the surface, cannot subside, and so we need to think about the direct effect of sunshine on the cloud temperature. It is true to say that most of the time the effect of heating by sunshine is quite negligible compared with other effects, of which the vertical motion is the most important. In fog the only vertical motion is the circulation in the cellular convection and this is the result of a fine balance between the loss of heat at the top by radiation and the gain from the slightly warmer sea.

In that situation the absorption of sunshine might become a determining factor. A ray of light from the sun might undergo as many as twenty deviations of its path by passage through droplets. Inside the cloud, light comes from all directions and about as much goes back out of the top as gets through to the surface below; the precise proportions depend very much on the depth and particle sizes within the cloud. But the path of the ray could be lengthened to twenty times the cloud depth, and this could make the solar heating important. It would cause some evaporation by raising the temperature a degree or two, in the case of a fog with no other large influences at work. But as soon as the cloud gets more tenuous and the sun's disc becomes discernible the absorption of sunshine is reduced to virtually zero, and the job of getting rid of the fog is never completed! As the sun goes down the radiative heat loss takes over control and the fog thickens up again until equilibrium with the sea temperature is re-established.

SHIP TRAILS

Ship trails are a rather rare phenomenon which was discovered by satellite. As far as we know they occur mostly in the eastern North Pacific and in the north and east of the North Atlantic Ocean. A survey of the North Atlantic for six years showed that they occur not very regularly and on the average about once every six weeks.

The trails occur only in a particular type of cloud. This is shallow, capped by a stable layer, with rather weak cellular convection, which appears to be caused in the same way as in ordinary layers of cloud except that the cloud is dull and full of small holes, and because the holes may be smaller than a pixel, the cloud's appearance is a mixture of the cloud and the surface beneath. This is not the full explanation, because even where the layer is complete it is still not as bright white as usual. It therefore seems that it is composed of a smaller number of droplets of larger than normal size, and this gives it a reduced albedo (the degree of whiteness). This implies that the air in which the cloud is formed has a deficiency of condensation nuclei and the cloud water has condensed on fewer but larger droplets.

The effluent from a ship's chimney contains a vast number of nuclei, probably many more than 20,000 in every cubic centimetre of exhaust. These nuclei are diffused (by the cellular motion already referred to) throughout the cells into which they are introduced by the ship, after the buoyancy of the hot gas has carried them up to the cloud top. The result is that the number of droplets is greatly increased and the cloud becomes bright white, like ordinary cloud, where it has been polluted.

We do not see trails in ordinary cloud because they already have a plenitude of condensation nuclei. We have not seen ship trails in air which has recently been over a continent: thus there have not been any in the Mediterranean or off the east coast of North America.

Where does the very clean air deficient in nuclei come from?

In the Pacific there are 'Ocean Deserts' in the middle of almost permanent anticyclones where there is no life in the surface water because of lack of nutrients. In these places such nuclei as may originally have been in the air have fallen out, and have not been replaced either by nuclei of biological origin or by particles of salt spray because the ocean is so persistently calm.

There are no such areas in the Atlantic and it appears most likely that the clean air comes from the frozen Arctic, mostly via the Davis Strait, but occasionally from the Greenland Sea, between Greenland and Spitzbergen. Two of the very rare occurences of trails in the North Sea were supplied that way. The Arctic air must travel slowly, for otherwise salt spray is produced; and it must be subsiding for otherwise the gusty winds of showers in such cold air would do the same.

While no measurements have yet been made in the clouds in which ship trails appear, the consistency of 'the story so far' gives confidence that future discoveries will not contradict it. Even so, we are sure there is much more to be discovered about the trails. The persistence of the trails, their narrowness many hours after formation, and their great length, tell us that the dispersion of pollution is sometimes – perhaps often – much slower than had previously been thought.

▲ 12.6 (Lac Genin, 18/8/61, 07:50 AM, N) This steaming lake lies in a mountain hollow in Burgundy. Cold air drains down the hillsides and is colder than the lake. This August picture is less than an hour after the morning sun reached the valley bottom.

▲ 12.7 (Cader range, 11/4/63, 19:35 PM, N) Cold air, characteristically to be found in 'frost' hollows, may be seen to cover a convex mountain slope when the wind is opposed by the katabatic (downslope) wind in the evening, even on a slope still in sunshine.

▲ 12.8 (14/4/68, 13:00 PM, S)   12.10 Tyrone, 8/5/68, 8:00 AM, NW) ▼

▲ 12.9 (Channel Islands, 13/7/67, N)   12.11 ▼

12.8 This ground is steaming in the late morning sunshine after stratus had deposited drizzle on it in coastal fog near Aberystwyth.

12.9 Thin fog in the Channel, near the island of Jersey, is too thin to absorb a significant amount of sunshine. The distant wave clouds are over the Cotentin peninsula about 55km (30 miles) away.

12.10 A pulp mill in Pennsylvania emits moisture-laden effluent with hygroscopic particulate pollution. Radiation sometimes converts this into a layer of stratus cloud, which then consolidates a valley full of cold air, possibly with smog.

12.11 An example of a cold pool produced at the site of 12.10 draining under the warmer air of the neighbouring valley.

◄ 12.12 (Jura, 6/10/62, 14:45 PM, E) A scene in Central Europe where an inversion below the mountain tops traps pollution and often cools enough to form cloud.

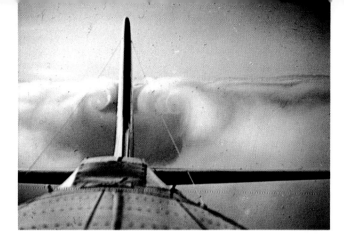

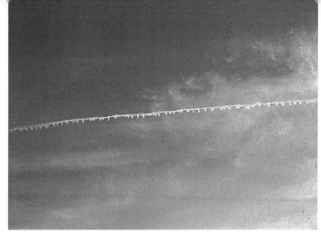

▲ 12.13 The downwash produced by the weight of an aircraft is here made visible by flying just above a layer of cloud. The vortices from the wingtips are beautifully revealed.

▲ 12.14 The downwash of a single-engined aircraft made visible by a contrail as it breaks up into blobs due to the stable stratification of the surrounding air which resists the downward motion.

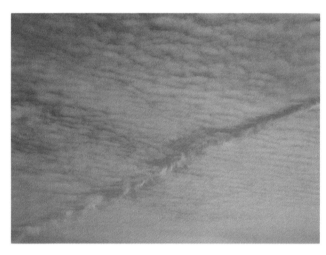

▲ 12.15 Just like a wing, a propeller blade also leaves a vortex behind in the track of its tip. The helix is distorted by the obstruction of the flow by the wing. The condensation occurs in the low pressure caused in the centre by the centrifugal forces around the vortex core.

▲ 12.17 A distrail may be caused in a thin cloud by aircraft downwash while glaciation may be started by the contact of the aircraft with some cloud particles, leaving a trail of fallout.

▼ 12.16 A vertical curtain of cloud which forms a contrail is drawn out into an almost horizontal sheet by wind shear in the air.

▼ 12.18 Occasionally a trail appears twisted, and the explanation is not known. This is a challenge to observers – to record the changes leading to this appearance.

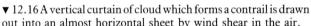
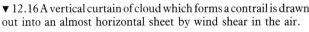

173

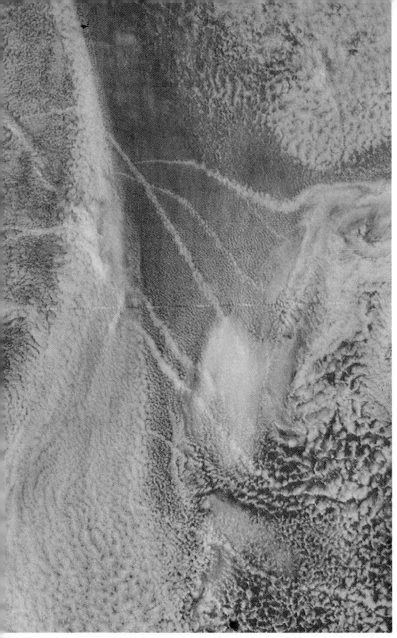

A most interesting feature of ship trails is that they persist for many hours and widen and disperse very slowly by comparison with air pollution over land. In some cases they have remained for as long as thirty hours. This carries an important message for those studying the transport of pollution across the sea: it means that the incineration of chemical waste at sea may produce quite high concentrations of obnoxious fumes which may reach the coast or pollute a passing ship. Pollution trapped beneath an inversion over land may travel, with very little dilution across a sea, and arrive at the other side almost in the same condition as when it first arrived over the sea at the opposite coast.

▲ 12.19 (17/8/83, 16:24 PM, 2)

12.19 Ship trails in the North Atlantic, at 60 deg N 32 deg W seen here are well to the north of the main shipping routes. This picture illustrates the typical cellular cloud which is made much whiter by the provision of a large number of condensation nuclei by passing ships.

The cloud cellular structure of droplets (not the amount of liquid water) is greatly increased. The width of this picture represents 500km (311 miles).

◄ 12.20 (8/2/84, 15:54 PM, 2)

12.20 Here we see copious ship trails on the much frequented lanes off the west coast of Portugal. The cellular structure of the cloud causes the trail to have a zig-zag appearance. Several trails can be seen in the making. The width of the picture is 750km (466 miles).

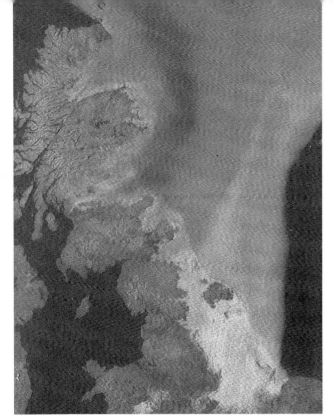

▲ 12.21a

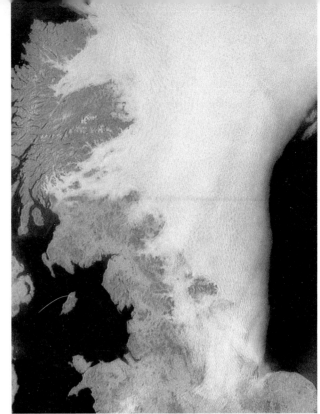

▲ 12.21b

12.22 ▼

12.21a (Channel 2), b (Channel 3) 27/4/83. North Sea stratus is a very common occurrence and the extent of its penetration inland into England and Scotland is always difficult to forecast. There are physical changes which occur when the sun shines to make the stratus over land whiter than over the sea, so that the coastline is clearly seen 'through' the fog. The greater strength of convection over the land causes the the top of the cloud to contain a much greater number of the smallest drop sizes, which give greater scatter by diffraction, while in Channel 3 particularly, much of the scattered radiation is absorbed by the greater proportion of larger droplets over the sea.

12.22 (17/9/82, 13:24 PM, 1) When the wind is blowing from the south it has been the practice for many years for chemical waste to be incinerated at sea (at about 55 deg N 4 deg E) because imperfect incineration might produce highly toxic effluent which would not be acceptable on land. However, this picture shows a plume from an incinerator ship visible to a distance of 120km (75 miles) from the ship, and no doubt detectable at a much greater distance.

The east coast of England and the west coast of Holland can be seen in the lower corners of the picture.

Quite dense industrial haze from the continent can be seen drifting in a curved path towards NE England.

The width of this picture is 250km (155 miles).

On some occasions the plume from an incinerator ship has been detectable by satellite for more than 300km (186 miles).

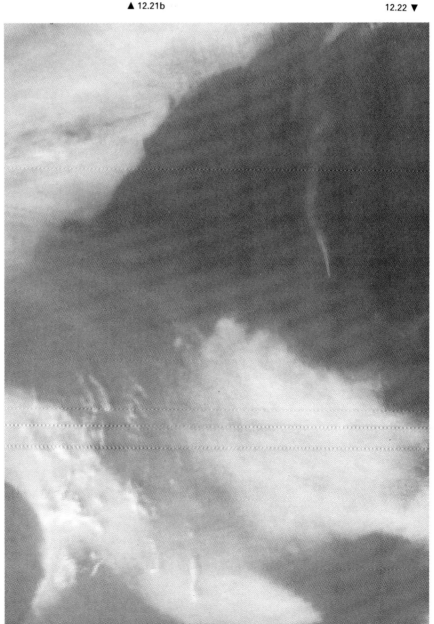

◄ 12.23 (19/8/64, 18:45 PM, NW) Sea salt haze is produced by breakers on the beach. If the beach slope is small and the wind strong the spray causes a significant reduction of visibility. This place is, significantly, called Saltburn-by-the-Sea, and is on the north-facing part of the Yorkshire coast of eastern England.

A gusty wind over the ocean soon blows up white-caps and generates a salt haze with an abundance of nuclei. Ship trails are not seen in the air where there have been gusty showers or strong winds, both of which generate condensation nuclei by the fragmentation of water film in the foam, and the bursting of bubbles at the surface.

12.24 It is very common for today's passenger aircraft to form contrails. This is a fairly typical scene from the Atlantic just west of Ireland. There are several navigation points which are traversed by many aircraft on adjacent routes. Some of these trails, for example those near the bottom of the picture, are spread out by shear as in 12.16. Since these trails are persistent having been formed in air just supersaturated for ice, what we see could have been generated over a period of several hours.

The fact that only a few show the effects of shear indicates that they are not all at the same height; but they are much colder (whiter, in this infra-red picture) than the clouds below, through which some strips of even darker (warmer) sea can be seen.

Since most of the airports of Western Europe have very few movements during normal sleeping hours, and this is 9 am by the sun, they have probably all been generated in the last three hours or so. Similar scenes may be seen in the middle of the afternoon and in several different areas of the North Atlantic Ocean. They may be found in places where new frontal cloud is about to form or where frontal clouds have evaporated and the air has subsequently ascended a little so that it has become supersaturated for ice.

▼ 12.24 (29.8.87, 09:01 AM, 4)

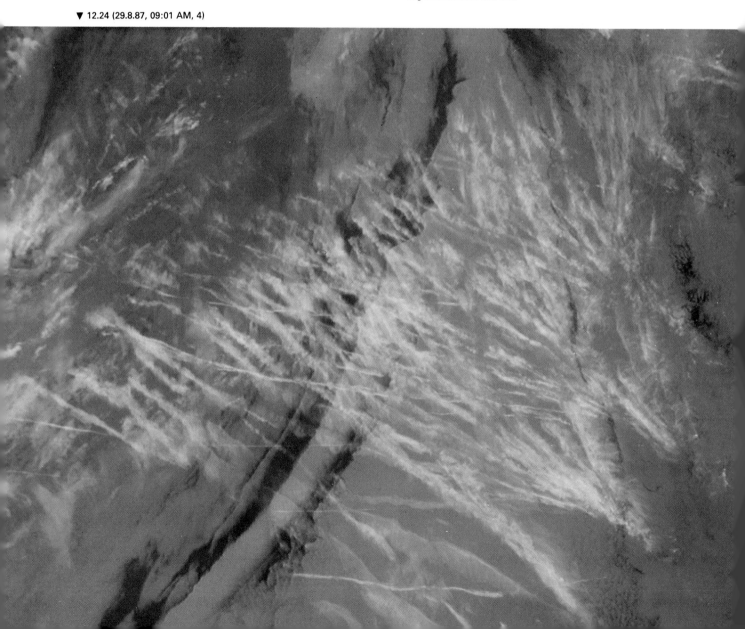

13
HABOOB EFFECTS IN THE ATMOSPHERE

In the atmosphere there are many situations where air of one density is brought into contact with air of another. All updrafts and downdrafts accomplish this, and produce a variety of small-scale turbulent motions at the mixing interface, as is best seen on cumulus tops. Rising or descending air eventually reaches a level at which the motion becomes horizontal or slows appreciably. The two places where this is most apparent is with cold air at the ground and warm updrafts spreading out as an anvil at the ceiling. Once the air begins to travel along the bounded surface, it assumes the characteristics of a gravity current with a particular shape and structure that is occasionally visible in the cloud formations.

The haboob, or dust storm of the Sudan, is an outstanding example of this process made visible by dust particles lifted off the arid ground by strong, cool surface winds, usually from a storm outflow. The interface is steeply inclined away from a 'nose' at which the cool air decelerates, rises sharply, then slides back above the advancing current. The nose proceeds at about 10–15m/sec (11–16.4yd/sec) but this is only about three-quarters the speed of the air behind it. The nose may project slightly forward just above the ground due to surface friction below it. There will also be lobes where cooler air pushes ahead discontinuously, trapping some of the clear air like the nodules on top of cumulus do. The forward bulge, or 'head' is the direct result of the deceleration since the denser air has nowhere to go but up and back around the outside of the faster flow within the structure.

The haboob is only a special example of advancing, cool, denser air. Others include sea breezes, cold fronts, and most commonly, the outflows from convective showers. In most cases, though, the effect is not visible because clouds rarely form that close to the ground. Only an occasional, short-lived piece of scud (see 6.20) or greatly lowered cloud base will reveal any clues to this complex and fascinating structure. The wake region is sometimes visible by inference as the high, irregular cloud base or outright clearing behind the leading edge of the outflow (see 6.24 and 6.39).

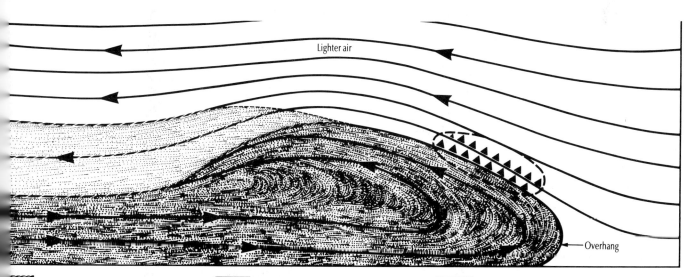

Lighter air

Overhang

Wake region of mixed air   Cold dense air, advancing   Region of maximum ascending motion

13.1 The typical form of the 'cold nose' at the front of a cold heavy air mass penetrating under a warmer one. It leaves behind a mixed intermediate layer due to turbulence beginning near the top of the nose.

The front does not advance uniformly as indicated in this cross-section, but develops lobes moving irregularly ahead as seen in 13.2.

The overhang at the front is due to friction at the bottom.

▲ 13.2

13.2 A model experiment by H. O. Anwar, in which a milky fluid, lighter than the clear water in the tank, was released near the top on one side. Here it is seen advancing across the top of the water as seen through the glass wall on the opposite side. The reflection in the top surface can be seen. The advance is in the form of 'lobes' characteristic of such a gravity current, and is like the spreading of the anvil of a cumulonimbus beneath a strong inversion which it does not have enough buoyancy to penetrate.

13.3 This storm, moving quickly southeast, has just begun to collapse. A powerful outflow surge is cutting into warmer air to form a set of nearly evenly spaced 'teeth', or lowerings (bright parts) below the dark shelf cloud base. They are tenuous masses of cloud material forming in the sharply rising, narrow mixing zone at the front of each lobe along the interface. The dark, adjacent spaces are free of low cloud because warmer air is diluting the outflow slightly as it becomes trapped between the lobes.

▼ 13.3 (Texas, 22/5/85, 8:25 PM CDT, W, 74deg)

13.4 Showers to the west are sending out a surge of cool air. This odd lowering formed briefly on the outflow boundary and could be seen rotating slowly clockwise. However, it was not clear whether the rotation was in the horizontal or vertical plane. Such features can extend to the ground as a gustnado when intense rotation is induced at the intersection of separate lobes of the outflow front. The spiral structure also suggests the kind of rolling motions expected on the upper surface of a gravity current head.

13.4 (Iowa, 1/6/87, 7:51 PM CDT, SW, 80deg) ▶

13.5 An explosive updraft pulse has hit the ceiling and spread out to form several 'knuckles' or lobes that are catching the sunlight. The return downdraft generated by the collapse of an overshooting top may be almost as important in creating an outward moving gravity current, and may explain sightings below the main anvil as in this case. Similar, earlier activity has left mamma-like remnants on the anvil slope above the bright lobes.

▼ 13.5 (Oklahoma, 2/6/85, 8:11 PM CDT, NNE, 74deg)

▲ 13.6a (Alberta, 27/7/81, 1:13 PM MDT, NW, 23deg)

The situation is somewhat different when an anvil spreads away from a cumulonimbus updraft because the 'surface' is flexible. The deceleration extends over a zone whose thickness and impact are dependent on the strength of the updraft and the temperature profile of the tropopause. The results are complicated by winds and wind shear, fallout and downdrafts, variations in density due to glaciation, and multiple updrafts or thermals but one factor remains common to every case; deceleration of the rising air causes lateral movement (and some turbulence effects), and this motion along a loose boundary induces some of the same buoyancy forces and lobe or mamma-like structures near the anvil edge as are seen in a typical gravity current.

13.6a–c Here are three examples of the haboob effect on the edges of thunderstorm anvils. In (c), a nose structure is evident. In (a), the cloud top (upper left) is splitting into 'toes' even though only a single, uniform tower had been seen rising a few minutes earlier.

▼ 13.6b (S Dakota, 21/6/83, 9 PM MDT, NNW, 13deg)

▼ 13.6c (New Mexico, 25/8/83, 7:34 PM MDT, NNW, 30deg)

(Texas, 25/5/87, 7:32 PM CDT, WSW, 46deg)

13.7a–c In (a) extremely rapid growth is occurring on the northwest flank of a storm moving to the north-northeast. The rain area is at left, and the orientation of its flanking line (west-northwest – east-southeast) suggests this is a left-mover type. The new towers are blasting upwards, then shearing to the right in a sculpted mass of folds and bulges that indicate outward squeezing of air which is nearing the stable ceiling.

(b) Is a blowup of a small section of (a). The sharp anvil edge indicates rapid outward motion, and since this is below an older anvil at the ceiling it suggests that the haboob effect can form below an expected boundary, as long as the deceleration is substantial there. The overall shape also hints at an outward and downward rolling motion.

In (c), outward motion is slowing and glaciation is softening the edges.

▲ 13.7a

▼ 13.7b

▼ 13.7c

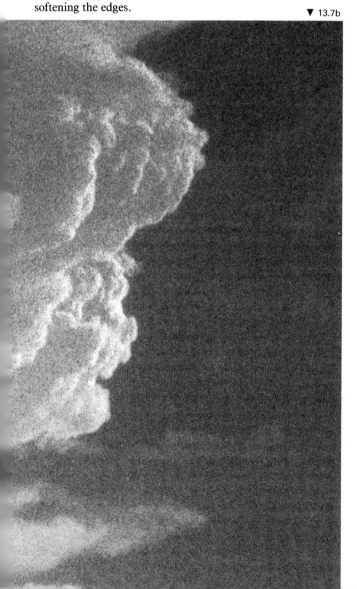

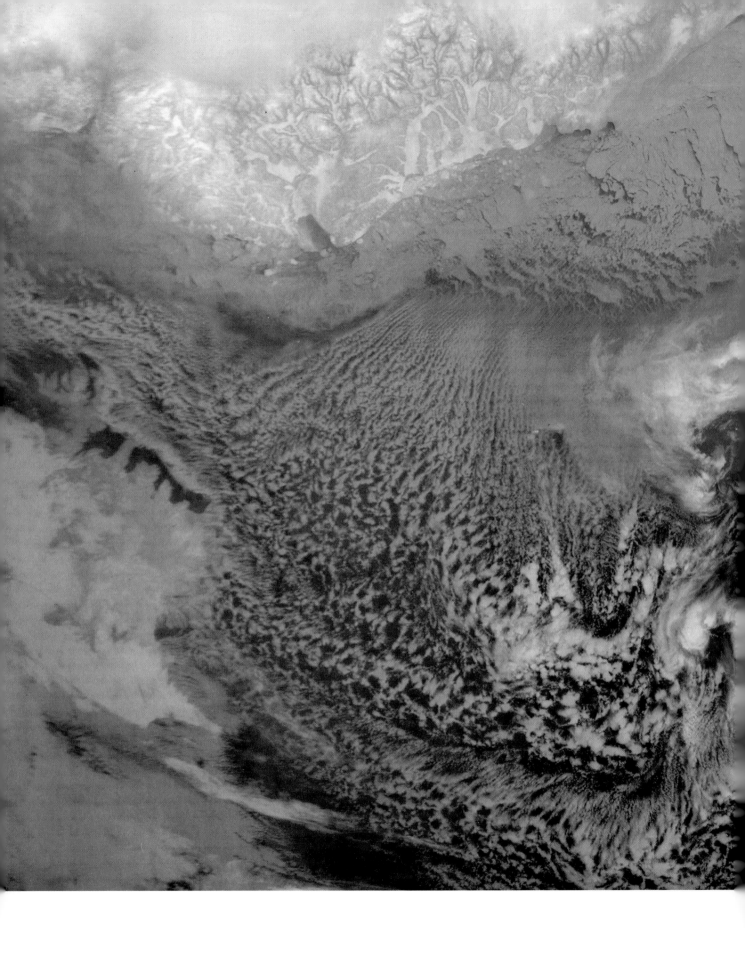

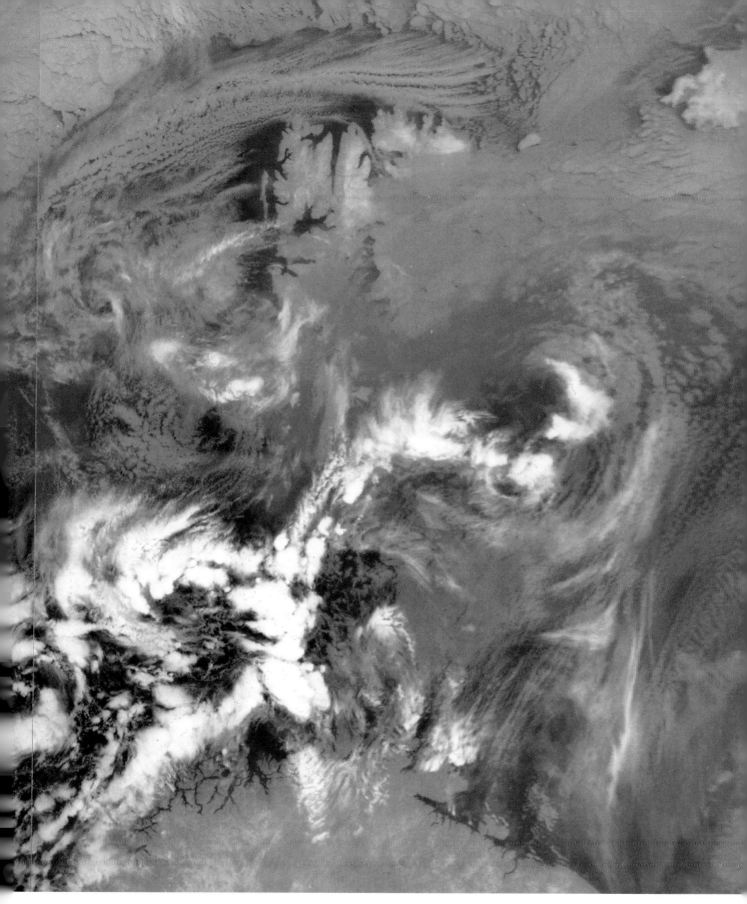

▲ 13.8 (28/1/85, 12:10 PM, 4)

In the darkness of the long winter night the satellite depicts the storms of the Norwegian Sea where the warmth of the Gulf Stream produces large cumulonimbus whose anvils are curved like the outflow at the top of hurricanes.

On the far side it is called the Greenland Sea and streamers of cloud indicate the outflow of arctic air across the frozen sea (top). On the left that air is spreading out towards Iceland (extreme left), but in the centre it is feeding the storms close to the coast of Norway (lower right). Spitzbergen and Franz Josef land are at upper right. The open sea appears black.

APPENDIX

THE PHYSICAL NATURE OF CLOUDS

We here apply the basic principles of classical physics to the study of clouds. The great names in this part of the human scientific endeavour begin with Archimedes, who clarified the concept of buoyancy. Among those who made significant contributions we find Boyle, Dalton, Joule, Clausius, Prévost, Planck, Maxwell, Kelvin and Napier Shaw. In the last three decades many scientists have contributed to the more complete understanding of meteorological mechanisms, but in the study of clouds, rain and hail the name of Frank Ludlam stands out pre-eminently. For him the visual study of clouds was a prime source of inspiration.

Clouds are composed of water droplets and ice crystals. We measure their sizes in microns (μm), a micron being a micro-metre. 1,000μm = 1mm (0.04in), and a million microns make 1 metre (3⅓ft). The droplets are so small that to them the air is a very viscous fluid, and they fall very slowly through it. They are therefore carried around by the air motion like smoke particles. Their fall-speed increases with size as indicated by Table I.

Table I. The fall-speed of water droplets in air.

Diameter in microns	1	10	100	200	300	1,000	2,000	3,000
Fall-speed, cm per sec	.03	3	27	72	102	400	650	800
		cloud		drizzle			rain	

3,000μm is 3mm (0.12in), and at that size raindrops have a fall-speed of 8m (26ft) per second. If they grow larger they are soon broken asunder by the aerodynamic forces into several smaller droplets. If we observe raindrops much larger than 3mm (0.12in) in diameter they are probably hailstones or snowflakes which have melted just before reaching the ground. All rain starts as tiny cloud droplets which are formed when the air is cooled to a temperature below its **dew point**. At that temperature the air is **saturated** and cannot hold any more water in vapour form. When the air is cooled further condensation takes place on **condensation nuclei** which are very plentiful in the air but they are very varied in size. Over the sea and near the coast there are usually numerous particles of sea salt originating from dried-out sea spray, and these are among the largest nuclei. In cities, and indeed near almost any human habitations, there are much larger numbers of much smaller nuclei which are mainly combustion products including smoke and acid particles, and in dry areas there are plenty of dust particles which may be mineral in origin or dry vegetable waste ground up by mineral dust. There are also nuclei of biological origin in the sea, and almost innumerable 'Aitken' nuclei which appear to be molecular in size and of very varied composition which can nucleate droplets as small as 0.1μm.

LARGE DROPS, SMALL DROPLETS

By far the most important mechanism which cools the air to form clouds is ascent. The pressure decreases upwards in the air as the weight of air above becomes less: consequently any parcel of air expands as it rises. In expanding it pushes aside the air surrounding it, and this uses up its internal energy so that it is cooled. The motion of the air molecules is its heat, and their bouncing off the surrounding air is its pressure; and some of their energy is used up as they push the surroundings back.

The excess water molecules then begin to aggregate on any nucleus they encounter, and because of the lower temperature they stay together as tiny droplets. But some nuclei, such as sea salt or acid particles from fires, are hygroscopic: this means that they hold on to the water by means of chemical molecular forces. These nuclei grow into bigger droplets in the first moments of cloud formation than the others.

The more rapid the cooling the greater the number of nuclei that are activated and become droplets. But another factor soon intervenes – surface tension, which is like a tight skin on the surface of a liquid. The most spectacular example of surface tension is that on liquid mercury, which enables drops to roll around like ball bearings. Drops of water behave in the same way on a dusty or greasy surface. The actual magnitude of the tension in the surface of the droplets is the same for all sizes. But inside a small droplet, whose surface has a great curvature, it produces a greater pressure than inside a large drop. This has the effect of squeezing the water molecules and making the smaller droplets more liable to evaporate. A hygroscopic nucleus has the opposite effect of holding on to the water molecules.

The molecules are escaping from the droplets all the time, while others are being captured from the surrounding vapour; so on balance, if there are many different nuclei present, as the rate of ascent slows or ceases altogether, the larger droplets will grow while the smallest will evaporate.

So, new clouds in rapidly ascending air contain large numbers of droplets including numerous small ones, while in slowly ascending air there are fewer small ones, and this distinction will be increased if the ascent ceases. If the cloudy air is mixed with surrounding clear air which is not saturated, any small droplets will be the first to evaporate, leaving only the larger ones which, having taken longer to grow, also take longer to evaporate.

CLOUD DENSITY AND DROPLET SIZES

The volume of a sphere of diameter d is $\frac{1}{2}\frac{\pi}{3}d^3 = \frac{1}{2}d^3$ roughly. Its cross-section area is $\frac{1}{4}\pi d^2 = \frac{3}{4}d^2$ roughly. If all the droplets have a diameter of $10\mu m$ ($= 10^{-3}$cm) the number in 1 cu m of air in which 1g of water is condensed as cloud (ie 1 cu cm, 0.06 cu in, of water) is $\frac{1}{\frac{1}{2}\times 10^{-9}}$ $= 5\times 10^9$. This is the same as 5,000 droplets per cu cm (0.06 cu in).

Each droplet has a cross-section area of $\frac{3}{4}\times 10^{-6}$ sq cm, and to obscure 1% of the area we require $\frac{4}{3}\times 10^4$ droplets and these are found in a volume with cross-section 1 sq cm and length $\frac{4}{3}\times 10^4/5,000$ cm. The view is 5% obscured in length 40/3 cm, and the vision is reduced to 10% by 6m (20ft) of cloud. Thus it would be barely possible for a passenger to see the wing tip from the cabin of an aircraft flying through this cloud. The vision is reduced to only 1% by 12m (39ft) of cloud.

In an old cloud in which all the droplets had diameter $100\mu m$, instead of $10\mu m$, the visibility would be 10 times as far, and would be reduced to 1% in 120m (390ft) with only 5 droplets per cu cm (0.06 cu in), but each having 100 times the cross-section area of the $10\mu m$ droplets. If the liquid water content were twice as much, ie 2g per cu m (0.07oz/35cu ft), the reduction of the vision to 1% would occur in 60m (200ft).

In a fog, ie cloud on the ground, the liquid content might be $\frac{1}{2}$g (0.02oz) or less, and then the visibility would be twice as much, 240m (800ft) or more. Sometimes it is possible to see the disc of the sun (or moon at night) through a fog or other cloud but still not too brightly to look at. That means that the cloud is close to the thickness for marginal vision. This is proportional to droplet size, and inversely proportional to liquid water content.

If the cooling is very rapid there is no time for the larger droplets to be formed and nearly all the very smallest nuclei are used. The droplet size may then be $0.1\mu m$ or even less. Yet such a cloud is sometimes formed on the upper surface of an aircraft wing where the pressure is much reduced; for that is how the wing is designed to produce lift. This part of it is then shrouded in cloud. The air passes in at the front of the cloud and out at the back of it 2m (6½ft) away in about $\frac{1}{100}$th of a second (at a speed of 400kt) so that the cloud is condensed and evaporated in that short time. Such ghostly clouds flash in and out of existence as the aircraft passes through regions of high and low vapour content near ground that is wet in patches or where cloud is evaporating and the mixing with the surrounding drier air is going on.

SCATTERING OF SUNSHINE

Dense clouds obscure the sun, their bases look very dark and direct rays cannot penetrate them. Every time a ray encounters a droplet it is deflected through an angle which depends on how close to the centre of the droplet it is aimed. A narrow beam of light is soon spread out and it only requires a few encounters for much of the light to be reflected back in the general direction of the sun, or sideways, and from those directions the cloud appears bright white.

Clouds with the smallest visibilities in them look whitest. These are the newly rising towers of cumulus, for in them the largest number of nuclei are being employed.

Old cloud, in which the smallest droplets have been evaporated and the efficient nuclei have grabbed all the water available and become large but few, allows more sunshine through. When viewed from the general direction of the sun it looks grey, or even purple as it evaporates. With the sun beyond it it often looks white and has a 'silver lining', and soon the sun's disc becomes discernible through it.

GROWTH OF DRIZZLE AND RAIN

It would take many hours to grow a raindrop out of a cloud droplet if the only mechanism growth were condensation of vapour which had been evaporated from the smaller droplets. But we can see from Table I that as soon as drops of different sizes begin to appear they have different fall-speeds. Collisions begin to take place and this results in coalescence. This second mechanism begins in earnest when some droplets with diameters as big as $20\mu m$ appear, and this can happen within 15–30 minutes of the first condensation. The cloud must therefore be big enough for its interior to be protected from evaporation by mixing with the surrounding air for this length of time. With an upcurrent of 5m (16ft) per second the cloud would have to be 6,000m (20,000ft) deep to provide 20 minutes of such protection; therefore it is probable that droplets already grown in the early stages of the cloud's existence are mixed into the later upcurrents which rise through the cloud. The cloud as an entity must have existed for 20 minutes or more, but individual **thermals**, bodies of rising warm air of smaller dimension than the cloud, may travel up through the cloud in only 5 or 10 minutes and start the rain production. But if the upcurrents in the small cloud are too strong they will carry all the nicely growing droplets out through the cloud top and evaporate them in the surrounding air.

The first rain to fall from a cloud may consist of droplets so small that they evaporate before reaching the ground. Evaporation is common for rain when the cloud base is very high, eg 3,000m (10,000ft), and the air below very dry, as is often the case in castellatus cloud. In the more common case where cloud is formed by convection from the ground we know more or less the humidity of the air below and the evaporation can be calculated. The result is set out in Table II.

Table II.   Distance of fall of drops from cloud base before evaporation.

Diameter at cloud base	<80μm	>400μm	>1mm
Distance of fall before complete evaporation	<1m	>100m	>1km

185

Thus, if a raindrop is to reach the ground from a cloud base at 1km (½ mile) it must start with a diameter of 1mm (0.04in) at least. Drops of diameter more than 0.4mm (0.02in) fall a significant distance and would be detected by a pilot flying beneath the cloud. From Table I we see that the fall-speed for this sized droplet is around 2m (6½ft) per second, and so they could easily be carried up to the top of a modest-sized cumulus in the course of the drops' growth. We may also note that drops greater than 0.3mm (0.01in) diameter would be likely to fall back into the cloud if they were in air mixing with outside air at the top of the cloud.

FREEZING OF DROPLETS

If ascent continues far enough the air becomes colder than 0 deg C. But the droplets do not immediately freeze into ice because some sort of nucleus is required to start the crystal growth. No suitable nuclei are normally present, and cooling may continue right down to -40 deg C with the droplets remaining unfrozen, ie supercooled. Beyond -30 deg C or so freezing is much more likely and the time taken for it to occur decreases until at -40 deg C and beyond the freezing is virtually instantaneous without any special nucleus. For practical purposes it is safe to assume that no cloud can remain unfrozen when colder than -40 deg C except in some very special cases where the drops contain a large nucleus of sulphuric acid, which is rare.

At around -20 deg C clouds can remain unfrozen for hours, but this is less likely the greater the number of larger droplets. Sometimes, therefore, freezing may begin after a few minutes at -20 deg C, especially in cumulus where the coalescence mechanism has already begun. Since the freezing may be started by a few larger drops, or even because of the presence of a few suitable nuclei a very interesting situation arises because of the presence together of ice and supercooled water in the same cloud.

When the molecules of water are bound together in a crystal structure they cannot escape as easily as from a liquid droplet, where the binding forces are weaker. A crystal therefore exerts a smaller vapour pressure. Alternatively we may say that in the presence of crystals the saturation mixing ratio is less than in the presence of water. Or we can say that, at temperatures below 0 deg C the dew point (above which water evaporates into the air) is **lower** (colder) than the **frost point** (above which ice evaporates, and below which ice grows by sublimation from the air).

But the frost point is not the point at which condensation begins in the air to form a cloud of ice particles. It is the point at which frost begins to form on solid surfaces such as trees and car roofs. To form an ice crystal in the air the temperature must be lowered to the dew point so that cloud droplets begin to form, and then the droplets must freeze. There are no nuclei in the air on which vapour can start to condense as ice. From time to time claims are made that ice crystals have been observed floating about in the air in very cold weather, but these could have originated as we have described at temperatures below -40 deg C, or they may have been blown off a solid surface or from the ground in cases when there is a lot of frost about. There is no evidence that such ice nuclei play any significant part in the evolution of clouds.

RAIN FROM ICE CRYSTALS

The freezing of cloud droplets is slow to begin in a cloud of supercooled droplets, and when it does begin the few ice crystals which form immediately begin to grow by condensing the vapour which is at the dew point and therefore supersaturated for ice. They grow much more rapidly than the droplets ever did; so much so that the removal of some of the vapour causes the droplets to begin to evaporate. The crystals soon become much larger than the droplets and as they fall more rapidly they capture droplets and grow by accretion. In a big cumulus cloud all this takes place so quickly that the growth produces hailstones. At the other extreme, where the rate of ascent is slow, more freezing occurs before the ice particles have developed a large fall-speed. Droplets which come into contact with the crystals freeze, but because the freezing releases some latent heat which has to be got rid of before the whole droplet can solidify, a shell of ice encloses a liquid centre. When this centre freezes, the ice occupies a greater volume and the ice casing is shattered, like a burst pipe. Spikules of ice are ejected and soon make contact with other droplets and cause them to freeze. By the repetition of this process the whole cloud becomes frozen as the last remaining droplets are evaporated. Ice crystals stick together or become entangled to form snow flakes.

This mechanism of snow formation is known by the name of its discoverers Bergeron and Findeisen. It is the origin of much rain, which starts as snow and melts when it falls down to the **freezing level**, which is where the temperature is 0 deg C. It is also called the **melting level**, which is the more appropriate name in this case. Air pilots call it the freezing level because above it they suffer the risk of supercooled cloud particles freezing on to their aircraft's wings, blocking engine air intakes, or otherwise destroying the aerodynamic properties of their aircraft by the build-up of rime.

FALLSTREAKS

Many turret clouds (castellatus) are formed far above the freezing level and ice appears quite soon after their first appearance. Ice crystals are often seen falling from their base. At temperatures warmer than -10 deg C such clouds would remain unfrozen. At colder temperatures, down to about -20 deg C, some of the droplets freeze and grow in 10–15 minutes large enough to fall out of the base as a streak of hair-like appearance called cirrus uncinus. The name **fallstreak** is preferred because it is descriptive of the mechanism which produces this cloud form, rather than a statement of its appearance only and in a foreign language.

At temperatures colder than -30 deg C the freezing of the whole cloud, once begun, is rapid, and the ice particles are more numerous and smaller and scarcely show any evidence of a significant fall-speed.

The shape of fallstreaks varies according to the variations of wind speed in the layers through which they fall, and also on the humidity in these layers which determines whether the particles will grow bigger or evaporate.

ENDURING CIRRUS

Ice crystals grow or evaporate according to whether the air is supersaturated or unsaturated for ice. They will only persist without either turning into fallstreaks by growing or vanishing by evaporation if the air is exactly at the frost point. This is achieved when the cloud is first condensed at or colder than -40 deg C. The droplets then all freeze immediately, and grow enough to condense all the excess vapour. They will then remain the same size and have a fall-speed of the order of 2m (6½ft) per minute, which is slow enough for them to remain in the same kind of air for a few hours.

Ice crystals do not stick together if they collide. If they are formed in a deep cloud in the presence of some water droplets they will grow into complicated crystal shapes with many dendrites (branches) which cause them to become entangled with each other to make snow flakes. When formed at a high altitude where the vapour content is much smaller than in warmer layers below, they never grow big enough to acquire different fall-speeds and do not agglomerate into snow. In this way they maintain a small fall-speed.

RADIATION AND LAYER CLOUDS

'Alto-' is the name given to clouds which are not cold enough to have the fibrous, cirrus-like, appearance, but are high enough to have been formed without the recent ascent of air from the Earth's surface. They are termed medium (middle-level) clouds by contrast with high (cirrus) and low (cumulus- and fog-derived) clouds. We now ask what mechanisms affect a thin layer of medium cloud.

The cloud top loses heat by radiation into space, from which it receives no radiation. The base receives more heat from the ground than it emits because its temperature is lower. The clear air above, which is transparent to this radiation, remains warmer than the cloud and the air below remains cooler than the cloud base. Thus there is a stable layer both above and below the cloud which inhibits mixing of the cloud with that air. Within the cloud the air becomes unstable and is continually overturning. This keeps renewing the cellular structure usually seen in these clouds, and if there is a difference of wind velocity between the top and the bottom of the layer the cells become lined up in **billows**, which are like rollers between the upper and lower layers.

Although there is no appreciable mixing between the cloud and the air above and below there is usually quite vigorous mixing at the edges where clear and cloudy air are side by side. As mixing takes place some of the cloud particles evaporate, and this cools the air, which increases its density, which causes new eddies, which cause further mixing, which leads to the evaporation of more cloud, and so on but at a gradually reduced rate as the neighbouring air becomes more humid.

EFFECT OF SUNSHINE ON CLOUDS

Most of the energy of sunshine is in the visible wavelengths from deep red through to violet. The energy of the ultra-violet part is absorbed in the stratosphere, and most of the energy in the near infra-red behaves just like the visible part. Large parts of the deeper infra-red are absorbed by the water vapour in the air, and this absorption is spread over a great depth of air. The cloud droplets are transparent to nearly all the energy of the visible and near infra-red and do not absorb significant energy from it. All they do is to scatter the sun's rays in all directions. The only way they lead to a significant heating of the air is by causing the rays to go zig-zagging through the cloudy region so that they travel a much greater distance through the water vapour than if they went straight through. Even so the energy thus absorbed is small compared with the energy gained or lost by condensation or evaporation due to the up and down motion in the cloud. We can therefore say that the effect of sunshine on the temperature in clouds is negligible for practical purposes. This is not at all the same as saying that clouds have no practical effect on sunshine, for they scatter a large fraction of it back into space and therefore significantly reduce the warming effect on the Earth. In this way they exert a very important effect in controlling the climate. They also have a big effect on the loss of heat from the Earth by radiation into space, and this too has an important role in the control of climate.

EARTH RADIATION

We have just noted the cooling effect of the radiation emitted from the top of a layer of medium cloud and the warming effect of the radiation from the ground on the cloud base. It was Prévost who first taught that everything emits radiation according to its temperature and that this is how a hot body warms a cooler one at a distance without having material contact with it: they are both emitting, but the cold one emits less and receives more until their temperatures are equal.

Gases and vapours are different from solids and liquids in that they do not emit radiation of all wavelengths appropriate to their temperature, but only in selected bands in that range. They also absorb radiation from other bodies in those same wavelengths. They are transparent to the other wavelengths. The air is virtually transparent to sunshine, which we can see for ourselves is transparent to visible wavelengths (light); but although the oxygen and nitrogen which make up 99% of the atmosphere are also transparent to the Earth's infra-red radiation, the water vapour and carbon dioxide are very absorbent to certain wavelength bands. The carbon dioxide is distributed through the whole atmosphere and causes the 'greenhouse effect', slowing down the escape of radiation from the ground to space: it makes the lower atmosphere warmer and the stratosphere colder than it would otherwise be, but this effect is fairly constant and has little effect on the variations of the weather.

Water vapour is much more absorbent of its particular wavelengths and is also concentrated much more in the lowest layers of air. The result is that there is a rapid exchange of heat between the ground and the water vapour in the lowest few metres of air. When the ground is warmed by sunshine, heat is rapidly absorbed by these layers so long as the temperature difference is not small, and they are warmed without conduction or convection currents. It is from this radiation-warmed layer that thermals are formed to carry the heat up into the higher layers by convection The temperature differences between the ground in sunshine and the air 2m (6½ft) above may be as much as 20 deg C on a calm sunny day. Equally on a calm clear night when the ground is losing heat to space in those wavelengths which water vapour does not absorb, it can become as much as 20 deg C colder than the air above, and the radiative exchange between them cools the air. When this air is cooled below its dew point, fog is formed in it. We can see with our eyes how shallow the cooled layer may be when it is very humid by the depth of the fog that is formed. Often it is only 2 or 3m (6–10ft) deep, which means that the cooled layer is of the order of 10m (33ft) deep, only the lowest part of it being cooled to the dew point. Once the fog is created it blocks the direct exchange between the air above and the ground below and tends to become more dense rather than deeper. It is deepened if a wind develops and stirs the air up so as to cool a greater depth. If the wind is stronger the depth of stirred air may be enough to reduce the cooling of the bottom layer and evaporate the fog.

The depth of the layer from which thermals are formed is several times as great as these shallow mist layers because while the radiative transfer is taking place convection on a small scale can initiate the release of a thermal on a larger scale. The air at 5, 15, and even 45m (16–150ft) is unstable because it is warmed more by the radiation at the lower levels, and only needs a push from below to start it rising. The air above the fog, by contrast, is very stably stratified.

THE BLUE PLANET

The Earth's atmosphere is the only one in the solar system that is cloudy with large areas free of cloud. The clouds dominate and create the atmosphere's motion. What of the other planets?

Mercury, like the moon, has virtually no atmosphere because its gravity is too small to retain it. Venus has ten times as much atmosphere as Earth with the consequence that it is almost red hot at the surface: this has prevented life from gaining a foothold. Life on Earth has been able to get rid of the vapours and gases with large molecules whose radiative effects must have made clouds very persistent; and replace them by the beautifully transparent oxygen and nitrogen. Mars has a much smaller atmosphere with very little water vapour so that its storms are dust storms.

The big planets, Jupiter, Saturn and Uranus have atmospheres more dense than that of Venus, while Neptune and Pluto are far too cold to have any interesting weather.

Earth, the blue planet, is blue because of the scattering of sunshine by the tiny particles and molecules in the air which have a size comparable with the wavelength of blue light. And just as the atmosphere looks blue from below, it also looks blue from above. Sometimes the dry areas of the land appear red through the blue sky when viewed downwards at an angle not too far from the vertical.

ACKNOWLEDGEMENTS

The authors would like to thank all those who have helped to make this book possible

Thanks are due to the unstinting help and enterprise of Peter Baylis and his staff in the Satellite Laboratory in the University of Dundee and to their sponsors, NERC, for facilitating the production of the selected satellite pictures which appear throughout the book of the European and North Atlantic area, as well as photograph 2.1. Additional satellite pictures were kindly provided by the Japan Meteorological Association (2.10 and 9.15); NOAA/NESDIS/SDSD, USA (6.4f, 6.13g and 8.5); and Atmospheric Environment Service, Canada (5.20c).

Many diagrams were copied from *Clouds of the World* and *Environmental Aerodynamics* by R. S. Scorer and from the works of Charles Doswell III, to whom a very special word of thanks is due for his taking time out to answer all Arjen Verkaik's impossible questions about severe storms, and to review Chapter 6. Much inspiration came from the late Frank Ludlam's *Clouds and Storms*.

Arjen Verkaik's relationship to weather is incomplete without considering the very important contributions made by his wife and partner, Jerrine Verkaik. They share all the trials and joys of each project or experience equally, and their combined efforts have evolved to preserve their uniqueness within a common understanding of the sky. During the writing of this book Jerrine was a sounding board for testing new ideas or ways to express them, offered patient encouragement during those hopeless spells of writer's block, and was forever there when Arjen needed to rediscover the essence of the task before him. Later in the project her extensive science editing experience was instrumental in preparing the entire final draft for the publisher.

Jerrine's assistance also permitted the authors to become friends over the several weeks of hectic discussions and strategising when R. S. Scorer visited the Verkaik home three times, and made possible their visit to the Scorer home in the final stages. Margaret Scorer has lived through this exciting period and supplied indispensible encouragement and companionship during the long period of delay and frustration from the first contemplation in 1982 of a successor book to *Clouds of the World* until, at long last, the contract was finalised.

Up to that moment the authors had only met for about forty minutes in 1984, and the determination to bring this about was sustained by the encouragement of colleagues in the Royal Meteorological Society, of whom Malcolm Walker deserves special mention in this respect. With such different origins and backgrounds our trans-Atlantic partnership would appear extremely unlikely; but having met we could only feel that the spacious sky would be the limit.

Richard Scorer and Arjen Verkaik

Photo Credits
The majority of the book's photographs were from the SKYART Collection of Arjen and Jerrine Verkaik, and include the following:
jacket cover pictures
pp 8–9
all pictures in Chapter 4 except 4.8, 4.17 and 4.22
all pictures in Chapter 5 except 5.2
all pictures in Chapter 6
9.14, 11.11, 12.1 and 12.3
all pictures in Chapter 13 except 13.2.

Richard Scorer provided the following:
all pictures in Chapter 2
all pictures in Chapter 3 except 3.12, 3.16 and 3.25
4.17, 4.22 and 5.2
all pictures in Chapter 7
all pictures in Chapter 8 except 8.3 and 8.4
all pictures in Chapter 9 except 9.5, 9.7, 9.8 and 9.14
all pictures in Chapter 10 except 10.2, 10.7 and 10.8
all pictures in Chapter 11 except 11.6, 11.9, 11.10 and 11.11
all pictures in Chapter 12 except 12.1, 12.3, 12.7, 12.9 and 12.13.

For their photographs taken from *Clouds of the World:*

Alistair Fraser 12.9, J. Frizzola and M. Rosen 11.6, Michael Garrod 3.16, Eigil Hesstvedt and colleagues 9.7, Harold Klieforth 9.5, Steve Hodge 11.9, Charles Hosler 12.7, Capt. L. A. Milner 3.25, Vernon G. Plank 12.13, C. Rey 10.2, Robert Simpson 3.12, the late Robert and Mrs Symons 9.8, and Betsy Woodward Proudfit 4.8. Additional pictures were provided by H. O. Anwar 13.2, Ryuji Kimura and Hidenori Yokogawa 10.8, Roger Smith 8.3 and 8.4, Walter Tape 10.7 and 11.10.

Diagrams taken from the works of Doswell and Lemon include the following:
Figs 5.9, 6.5, 6.9, 6.11a and c, and 6.16 from Doswell, Charles A. III, 'The Operational Meteorology of Convective Weather, Vol. 2: Storm Scale Analysis', *NOAA Technical Memorandum #ERL ESG-15* (US Dept. of Commerce, Boulder, Colorado, 1985).
Figs 6.11b and 6.12 from Lemon, L. R. and Doswell, C. A. III, 'Severe thunderstorm evolution and mesocyclone structure as related to tornadogenesis', *Monthly Weather Review*, v107, pp 1184–97 (1979).

INDEX

Page numbers in **bold type** denote main references